Theory of the Object

Theory of the Object

Thomas Nail

EDINBURGH
University Press

Edinburgh University Press is one of the leading university presses in the UK. We publish academic books and journals in our selected subject areas across the humanities and social sciences, combining cutting-edge scholarship with high editorial and production values to produce academic works of lasting importance. For more information visit our website: edinburghuniversitypress.com

Edinburgh University Press Ltd
The Tun – Holyrood Road, 12(2f) Jackson's Entry, Edinburgh EH8 8PJ

Typeset in 11/13 Bembo by
IDSUK (DataConnection) Ltd, and
printed and bound by CPI Group (UK) Ltd, Croydon, CR0 4YY

A CIP record for this book is available from the British Library

ISBN 978 1 4744 8792 4 (hardback)
ISBN 978 1 4744 8795 5 (webready PDF)
ISBN 978 1 4744 8793 1 (paperback)
ISBN 978 1 4744 8794 8 (epub)

Contents

Figures

Acknowledgements

I am indebted to several people for their support and encouragement of this project. I greatly value my late-night philosophy strolls around Harvard Gulch park with Chris Gamble and Josh Hanan, where we discussed many aspects of this book at length and over several years.

I am grateful to my research assistants who helped me with various aspects of this project: Adam Loch, Kevin Buskager and Jacob Tucker.

I thank my anonymous peer reviewers for their constructive feedback and everyone I worked with at Edinburgh University Press, especially Carol Macdonald, for guiding this book to publication. I would also like to thank my friend and copy-editor, Dan Thomas, who passed away unexpectedly this year, for his excellent close corrections and suggestions, in addition to the copy-editing work of Andrew Kirk.

I also benefited greatly from line-by-line feedback on various chapters from Michael J. Ardoline, Marco J. Nathan and Katherine Robert. Michael Epperson was also kind enough to give solicited feedback on a related essay Chris Gamble and I co-wrote for the journal *Rhizomes* called 'Blackhole Materialism', related to several themes in Part III of this book. I learned about the exciting phenomenon of 'quantum scarring' from Diana Bull and Raoni Padui's New Year's card, of all places. Thanks, you two.

Thanks also to the University of Denver for its financial support for parts of this book's publication.

Finally, I would like to thank my family, and especially my wife, Katie, for her continued support and direction on parts of this project. These are the material conditions without which this work would not have been possible.

Introduction:
A World of Objects

We live in an age of objects. Today there are more objects and more kinds of objects than ever before in human history, and they continue to multiply at a mind-boggling rate. The sciences have succeeded in transforming almost every dimension of reality into one kind of object or another. They have mapped and catalogued nearly every corner of the earth. They have made commodities out of things that previous generations would never have thought to commodify, such as genetic codes, water, air, seeds and social care. Technological innovations now allow us to transcode almost anything into digital objects made of ones and zeroes. The scope of what constitutes an 'object' today seems to be unlimited. In this way, objects seem to have become synonymous with the nature of reality itself.

Yet our age is also one of flux. Everything we thought was stable, from subatomic particles to the cosmos at large, has turned out to be *in motion*. From the acceleration of the universe to the fluctuation of quantum fields, nothing in nature is static. The universe is expanding in every direction at an accelerating rate. What Einstein once thought was an immobile finite universe has turned out to be an increasingly *mobile* one.[1] This acceleration also means that even space and time are not a priori structures, as we once thought, but continually emerging *processes* in an unfolding universe.[2]

Even at the smallest levels of reality, what we previously thought were solid bodies and inviolable elementary particles, physicists now believe to be emergent features of vibrating quantum fields.[3] These fields' movement can no longer be understood as 'motion through space, evolving in time' as Newton had once understood it. Where an object is and how it is moving can no longer be determined at the same time with certainty.[4] The old paradigm of a static cosmos built from static particles is dead. All of nature is in perpetual flux.

How does it change our understanding of objects to know that they are composed of such *processes*? For example, quantum fields are not objects in any classical or relativistic sense. They are not in a single place, nor do they have any fixed properties. We cannot even observe them directly. We only record the *traces* of their interactions on measuring instruments. It seems that beneath the observable level of all objects are processes that *entangle* with our own with invisible threads.[5] If we take this seriously, as I think we should, our whole conception of objects must change.

Even the lowest energy state of quantum fields is not zero or static but fluctuates indeterminately – neither in one state nor another.[6] The nature of these quantum 'vacuum' fluctuations is perhaps the single greatest mystery and challenge facing cosmology and fundamental physics today. These fluctuations are nestled deep in the heart of every object, including in the dark energy that is accelerating our universe.[7] These indeterminate quantum fluctuations are the invisible *engines of nature*, and it is time we took them seriously, philosophically.

Although we live in a world saturated with objects, the most revolutionary discovery of contemporary physics is that *the foundation of objectivity is not an object at all but a process*. This fascinating and odd state of affairs is the motivation for this book – the aim of which is to offer a *kinetic process theory of objects*.

A new theory of objects

This is a book about objects and the history of their emergence. It is a book of theory but also takes the history of scientific knowledge and the creation of new objects seriously. I am, in particular, inspired by recent developments in contemporary science and maths that point towards a more process-based understanding of objects. This book is an attempt to show some of the broader conceptual consequences of these findings.

The existence of indeterminate quantum fluctuations, for instance, if true, means that there are no static essences, forms or substances in nature. However, since stable objects have been the focus of study in the history of Western science and philosophy, this raises an interesting question. What has Western science, maths, logic and technology been studying if not discrete objects? Furthermore, if scientific observations have always affected their objects, how does this change our understanding of objectivity? What is an object if it is continuously modified or transformed by other objects and subjects? The prime motivation of this book is to take this observer-dependence seriously and explore the hidden *processes* and performances that go into creating and sustaining objects.[8] It is a process philosophy of science.

However, although some quantum physics interpretations inspire me, this book is not a quantum theory of objects. I am not trying to explain all objects by quantum or any other kind of science. I don't think that is possible or desirable. I am also not trying to put forward an ahistorical or universal theory of objects. Philosophy is not the queen of the sciences. Theory is always situated somewhere and at some time with a particular vantage. Recent sciences or philosophies do not give us unmediated access to the unchanging essence of objects in themselves. They are more like starting points for a philosophy of the present, not a conclusion.

As a work of philosophy, this book aims to introduce a few synthetic concepts to help us hold together and make sense of a wide range of diverse practices of object-making. The book's ideas are meant to synthesise a big picture of what we are doing when we define, measure, order and talk about objects.

Specialists in the sciences will no doubt find that discussions of their particular field go by too quickly or are too cursory. Historians may find the sheer amount of time and geography covered and organised into four types of objects to be too broad a stroke to be of use. Philosophers may find too much history for their tastes and prefer to stick to the more conceptual chapters. These are valid disciplinary concerns.

However, the strength of this book lies in its synthetic and systematic nature. I am not doing historical analyses as such, nor pure philosophy, but rather marshalling the historical material to generate broader conceptual conclusions. This book is for readers interested in drawing connections across history, philosophy and science, looking for a bigger picture than each discipline usually offers on its own.

At the broadest level of inquiry, I want to know more about the general conditions for the emergence of objects. What makes possible the existence and persistence of a mundane object like the ketchup bottle I hold in my hand? As we look closely, I think we find that this bottle cannot be separated from the *history* of how objects get made more widely. In this book, I try to show that the past is not gone but has secretly rolled itself up inside the present like a flower's folded petals. I do not mean that the past is relevant because it leads to the present, but that past patterns are really and truly still active in the present. We will not find these patterns by looking at ketchup under a microscope. Nor will we understand the bottle by looking at the creation of *any particular* container of ketchup. We need a deeper historical account of object patterns immanent to the bottle.

Suppose physics and cosmology are correct, and there is nothing in nature that is not in motion and process. In that case, we should seriously reconsider what this means for our theories of objects more generally. If nothing is static or entirely independent of observation, then what are the broader interpretive consequences for understanding the history of science and technology, where scholars and scientists have often assumed the opposite?

Beyond science and technology, of course, the whole world is also made of objects. Humans are not the only ones who make them. In the most general sense, an object is just a metastable formation of matter in motion. At least in my view it is. An object is an eddy in cosmic energy dissipation – a temporary layover on its way into the cold darkness of space. However, this definition covers such a vast range of objects that I cannot cover it historically in a single book. *Theory of the Object* focuses on the proliferation of objects in human scientific history.[9]

I should also note that this book is part of a larger philosophical project to rethink politics, ontology, art, nature and science from the perspective of *movement*.[10] I find it endlessly fascinating that something as simple as movement has posed such enormous difficulties for philosophers and scientists in the Western tradition. Why have the greatest minds of Western civilisation dedicated their lives to discovering something genuinely immobile? Aristotle's idea of an 'unmoved mover', Archimedes' fixed 'point', Descartes' 'unmovable' certainty, Newton's divine clockmaker, and even Einstein's idea of a block universe are all part of this grand effort. But what motivated this critical pursuit, and what are the consequences of it for us today? This book is one angle on answering this question. Specifically, it is interested in how scientific and mathematical knowledge has responded to this question with objects.

However, I do not mean for this book to be a complete history of science or every great scientist in the Western tradition. *Theory of the Object*'s unique contribution is that it offers a movement-oriented process interpretation of the history of science and the formation of objects. This is an appealing perspective because science and mathematics have tended to treat objects as discrete and stable. Movement is typically the thing that is explained by something else. Why? And what would a history of scientific objects look like that did not reduce motion to something else? I would like to offer a different, but by no means unabridged, account of motion in science. My focus on movement is an essential aspect of my account's novelty and I hope sheds a helpful light on many difficult issues.

More specifically, I have identified four significant patterns of motion that characterise object formation in history. Each pattern is associated with a large block of history (prehistory, ancient, medieval/early modern and modern). These are not necessary, universal or developmental periods or patterns, nor do they fit neatly into a linear schema. They all mix through history, but to varying degrees. Tracking that mixture is a huge task, so in this book I focus mainly on the dominant pattern of objects during each historical period.

I would also like to stress that the theory of the object I am proposing emerges *historically* and does not come from a priori categories. This means that it is not only objects that are kinetic processes but that the history of the theory of objects, for me, is also a kinetic process. This move means that this book can only show its theory by performing it through history. The book is itself a historical object that contains the hybrid history of the objects it describes.

Of course, scientific practice has political, aesthetic and ontological dimensions and influences. They cannot be separated, and many great works have shown their entanglement.[11] However, by focusing on their entanglement, one tends to lose, at least in a single book, historical breadth. Here, as in my other books on the philosophy of movement, I have chosen historical breadth over the entanglement of domains.[12] I plan to write a more synthetic book in the future that shows the entanglement of the political, aesthetic, ontological and scientific dimensions of these historical patterns.

In this book, though, my focus will be on the sciences as practices of object creation. This is not all that science does, but is one aspect or way of understanding its primary activity. The two most common methods of interpreting the history of science are that scientists 'discover' pre-existing forms or merely make use of objects instrumentally. Instead, this book offers a different angle in which science co-creates and orders objects *with* nature. It is an immanent, performative and non-anthropocentric process philosophy of science.

I offer here, then, a 'theory' of objects in the sense of the Greek word *theōría* (θεωρία), as a 'movement, sending, or process'. *Theory of the Object* is a process of describing processes of movement. But let's look more closely at what movement means, what is genuinely novel about it as a theoretical lens, and what two main problems it can help us overcome in existing theories of objects.

Overcoming the idea of static objects

The first problem I hope my movement-oriented approach can overcome is the idea that objects are static things. There are a few versions of this 'static theory of the object', each with problems, which I think are worth tackling. Let's first look at these static theories, then consider their issues and see if there is a different way forward.

Objectivism

The first, and perhaps most pervasive, static theory of the object to look at is called 'objectivism'. In this theory, an object is a static, self-enclosed block of spacetime, which pre-exists its discovery by humans. For the objectivist, objects lie in wait, possibly unchanged for millions of years, until they are 'discovered' by humans. The actions, tools and conditions of the discovery do not change anything about the object. According to this view, the scientist's activities, environment and tools might help or hinder a clear picture of the object, but they are not part of the object.

In this theory, the object of scientific inquiry does not affect the instruments of observation, the environment or the observer. The object remains just as unmoved or unchanged as the observer in the discovery. Our theories *about* the object might change over time, or the object might come to *mean* different things to humans. Yet the object remains essentially unchanged no matter what we think about it. In this view, the immutability of the object is precisely what makes empirical verification and falsification possible.

Language, for example, has objective meaning only insofar as it can be verified accurately by a stable empirical object. If the object moves or changes the observer, the whole correlation between language and object suddenly becomes unstable – sliding on kinetic sand. It is no coincidence that historians of science like to tell developmental, evolutionary or progressive stories about our increasing knowledge over a fixed set of natural objects. The idea of progress in science and other fields of knowledge relies on an unchanging domain of objects that can be incrementally verified or falsified.

An extreme example of this kind of objectivism is that nature consists of geometric and unchanging mathematical forms. Since these forms remain unchanged by knowing them, objectivist mathematics and logic claim to have discovered a static and eternal 'language of nature'.[13]

Constructivism

The second static theory of the object I want to look at is called 'constructivism'. In this theory, an object is a *fixed mental state of an observer or group of observers*. Instead of progress through trial and error, the constructivist thinks that any correspondence between what a subject thinks about an object and the object itself is entirely arbitrary.

If an object is something that occurs only in the observer's fixed ideas or concepts, then there can be no basis for knowledge about the actual features of the object *in itself*. Whatever the object is, the constructivist says that all we have access to are *our* thoughts, words and actions.

For example, when a scientist observes a state of affairs, the constructivist points out that she cannot help but bring all her cultural, historical, linguistic, instrumental and institutional biases. These cultural beliefs are utterly incidental to the thing in itself. Therefore, the observer can never encounter the object in itself because the encounter and the observer's presence shape the way the object is experienced. So, for the constructivist, scientific objects are always local, contextual, collective, social fabrications fixed in human structures and have no knowable existence apart from the scientist's relatively stable observational context.

Again, it is no coincidence that historians tend to write the history of science from the perspective of scientific *ideas*, *thinkers* and *texts*. Constructivists tend to privilege what humans *think* about objects over analysing the changing material conditions that made their knowledge possible, including other objects and non-human processes.

Before we move on to the problems with these first two static theories, let's look at two recent alternatives to them. One way to avoid the problem of trying to get the object *in itself* to match up with the object *for us* is by not introducing the division in the first place. What if everything was only objects? We could then think of the subject as a highly composite type of object. So, the next two theories, instead of dividing the world up into subjects and objects, divide it up into *objects* and their *relations*.

Relational ontology

One version of this approach is called 'relational ontology'. In this view, an object is nothing other than the set of all its relations with other objects. In one popular version of this theory called Actor-Network Theory, relationships are primary, and objects emerge as nodes from pre-existing networks. Objects are what they *do* or how they *act* through their distributed systems. In a relational ontology, there is no such thing as an unrelated object.

Furthermore, in this view, there is no pre-given hierarchy among objects. Relations can always shift around and become different. Objects have no static essences because it is the broader network that defines and differentiates them. Objects are born and die, but network patterns, as such, do not because they *precede* and *exceed* all objects. If this is the case, though, what is the source of the change and novelty of the networks if it is not the objects themselves? How can networks change without objects that move?

One answer to this question comes from another kind of relational ontology called 'vitalist new materialism'.[14] In this view, relations are 'vital', 'virtual' 'forces' that create 'changes' in relations without any material movement in the objects.[15] Its proponents do not call this a 'static' view of objects, but in my view it still erases motion or kinetic change in favour of a virtual or relational change.[16] The French philosopher and founder of Actor-Network Theory, Bruno Latour, for example, rejects the 'static' view of objects, but instead proposes to replace it with a theory of 'successive freeze-frames that could at last document the continuous flow that a building always is'.[17] Latour explicitly accepts the label of 'secular occasionalist',[18] meaning that the world is made of discrete freeze-frames run together through time.[19]

But before getting into the differences with my approach, let's look at one final non-movement-based theory of the object.

Object-oriented ontology

This last theory is the most recent and the most different from the previous three. In object-oriented ontology, everything is objects and relations. Like the relational view of objects, this view agrees that objects connect in networks of changing relations. However, for object-oriented ontologists, objects are not reducible to their relations. Objects are 'discrete', 'stable', 'unknowable' 'things-in-themselves' with 'definite boundaries and cut-off points'.[20] Each object is 'vacuum-sealed' off from others and contains within it a secret or 'withdrawn essence'[21] that is 'singular' to it alone.[22] Graham Harman, a founder and proponent of this theory, describes it as a kind of Kantianism without a subject – everything is an unknowable object in-itself.[23]

Harman disagrees with the objectivists because he says they 'undermine' objects by reducing them to what they are made of (matter and particles). He also rejects the constructivist and relational views because he says they 'overmine' objects by reducing them to their network of relations. The typical explanation given by the sciences, he says, 'duomines' objects by claiming that they are just components of larger objects, which also have components.

The object ontologist worries that by defining the object purely by its relations with others, we explain the object itself by *something else*. We lose the 'real essence' of the object in-itself behind the appearance of its fluctuating relations. In this view, the only way to protect the object is to 'vacuum-seal' the 'unexpressed reservoir' of the object off from all its relations with others.[24]

How does this theory account for changes in objects? Harman splits the object into two parts. One part changes along with its relations, while the other part has 'hidden volcanic energy that could . . . lead it to turn into something different'.[25] This is why Harman criticises relational ontologies for not being able to account for change. 'Unless the thing holds something in reserve behind its current relations, nothing would ever change', Harman says.[26] In this view, the object's essence is the source of all change and motion, but only now and then. 'Stability is the norm'[27] because objects are mostly 'aloof [and] do not act at all: they simply exist, too non-relational to engage in any activity whatsoever'.[28]

Harman also claims that the essences of objects do not have 'an eternal character',[29] and can be 'transient'.[30] He accuses relational theories of being 'static' for the reason I said above.[31] Despite his protestations, however, he ultimately admits that the hidden parts of objects 'transcend' the world and do not engage in *any* activity whatsoever.[32] Since movement, as I understand it, requires activity of some kind, object-oriented ontology's ultimate position is one of immobility and stasis. So even though Harman *says* that change can come from something that has 'no action whatsoever', such a metaphysical belief amounts to a violation of every known law of physics.

Moving objects

What then is the philosophy of movement, and how does it offer a new way forward that overcomes the previous four theories? The philosophy of movement is a kind of process philosophy. This means that instead of treating objects as static forms, it treats them as metastable processes. Some movements are small and allow the object to remain relatively stable, like a river eddy. Other movements are more dramatic and can either destroy or transform objects, like a turbulent rainstorm.

By contrast, the four theories above define the object as static to some degree. As such, they are unable to theorise the movement, novel transformation and emergence of objects. Let's look quickly at the limits of each of these four theories and then see how my approach compares.

The problem with objectivism is that it treats objects as if they were unchanged by their discovery and observation. This view ignores the history, relations and agency of objects and treats them as entirely passive. But if objects are merely passive, how can they affect the other objects and observers that interact with them?

On the other hand, the problem with constructivism is that if the object is nothing other than what humans think or say about it, it is also robbed of all its agency and capacity to affect others. If objects are incapable of their own movement and novelty, then how do they emerge and change? Constructivism posits a radical difference between human subjects and natural objects that leaves humans trapped in their own world.

Relational theories of objects reject this division and acknowledge that objects act through their relations. The problem, however, is that these relations, by preceding and exceeding objects, determine them fully. Where then is the agency and motion of the object? How can the object introduce novel and generative motions into such relations? For Latour, the relations that constitute objects are, by definition, wholly determinate and mappable. Changes in relations do not originate from the movement of objects or their indeterminate materiality. Change occurs like a series of sudden 'freeze-frames' in the networks.

Finally, although object-oriented ontology tries not to reduce objects to unchanging essences, social constructions or relations, it saves the object only by completely sacrificing it. In the end, the object's essence ultimately transcends the world and is cut off from any relation to it. The core contradiction is that the essence of objects is the source of all change but does not act or move in any way. It is, therefore, ultimately, a philosophy of immobility and static transformation.[33]

These four theories of the object could not be more different. Yet they all try to explain the object's movement by something that *does not move* (an essence, a mental/social representation, a flat relationality, or an utterly inactive essence). The problem here is that all four theories *start with a division* either between subject and object or between object and relation.

What is different about the philosophy of movement? Instead of trying to explain movement by something else, the critical difference is that it starts from the historical knowledge that everything is in motion. I could be wrong about this. If proven wrong, I am prepared to concede my position and explore the new philosophical consequences.

From this perspective, I agree with Harman that objects are singular. We cannot reduce them to their determinate parts or relations. However, for me, this is because the movements of matter that comprise objects are not fundamentally determinate. Matter, or what physicists would more precisely call 'energy', at its smallest level, is 'indeterminate fluctuations'. These fluctuations are not particles, substances or objects, and we cannot directly observe or know them. Saying that objects are 'reducible' to indeterminate energy makes no sense. There is no determinate 'something' that is at the heart of the reduction.

Movement is indeterminate and relations are indeterminate relations. The movement of matter, in my view, has no higher or exterior causal explanation, or at least there is no experimentally verified one or hint of one yet. That is not to say that there aren't theories that try to interpret it away or that I cannot be wrong.[34] However, at the moment, I am putting my philosophical wager behind the real possibility that ongoing indeterminate movement is a fundamental feature of nature that we will have to live with. After Lucretius put the indeterminate swerve of matter at the heart of his philosophy in the first century BCE, commentators balked for centuries, but now this idea has returned with force. I intend to revive this tradition.

How might an object-oriented ontologist respond to my view? Graham Harman has already replied to the idea of quantum indeterminacy in a recent article on the work of the physicist Karen Barad. There he writes that 'undermining treats individual objects as too shallow to be the truth and seeks to replace them either with a micro-army of tinier things or a primordial lump of indeterminate flux'.[35] In response to this,

I would say two things. First of all, for Barad and myself, objects are just as 'true' as fluctuating fields, and the idea of replacing one with the other makes no sense, since objects *are* metastable fields. Second, there could not be anything less like a 'primordial lump' in the entire universe than indeterminate flux. One of the most pivotal events in the history of science was discovering that matter/energy is *not a substance* and has *no fixed a priori properties*. Lumps are undifferentiated, but indeterminate fluctuations are the processes of differentiation that create and sustain all differences. Harman, in my opinion, does not understand quantum indeterminacy and flux and this invalidates his objection.

It may sound like a small shift in starting points to go from stasis to movement, but it makes a huge difference. Each of these four theories has a method that follows uniquely from its starting assumption, as does my movement-oriented view of objects. Therefore, if we want a theory of objects that can make sense of their movement, emergence and novelty, these first four options will not work.

Instead of starting with stasis and trying to explain movement and process, *Theory of the Object* inverts this logic. It begins from the historical discovery of quantum flux and then tries to explain the emergence of stable scientific knowledge.

Theory of the Object also offers a new kind of process philosophy distinct from older models of process based on vital forces, as in the philosophy of Henri Bergson, or on static, discrete, strobe-like 'occasions', as described by Alfred North Whitehead.[36] My term for this third kind of process philosophy is 'process materialism' or 'kinetic materialism'.

If an object is not an essence, idea or relation, then what is it, according to a process philosophy of movement? We need to look no further than the kinetic origins of the word 'object', from the Latin *ob-* ('against') + *iaci* ('I throw'). The object is a fundamentally kinetic process. It is something thrown into motion and turned against or looped around itself. It is a fold. Instead of a discrete, vacuum-sealed atom, objects are continuous processes that fold back over themselves, making more complicated knots. My point is not that we have to accept this definition because the Romans used this word. The Latin roots indicate that there is some precedent for thinking of objects as processes that I believe are worth recovering. The object is not a discrete or static block of space and time, but a kinetic *process*.

Overcoming the idea of ahistorical objects

The second problem this book aims to redress is the ahistorical treatment of objects. If we define objects as having pre-existing static essences, as in the 'objectivist' view, then they have no real history because they have no real change. The *essential* laws and properties of nature always remain the same. In this view, the task of knowledge is to develop an increasingly accurate system of coherent and verifiable information through falsification. Progress is possible, in this view, because the laws and objects of nature have *no history*. The quest for universal laws of nature seeks this kind of ahistorical object.

On the other hand, if we adopt the 'constructivist' view, the contents of objects change, and so it seems as though there is a real history of objects. However, as much as the meaning of things changes *for us*, their conditions remain static ones determined by the ahistorical structure of human reason.[37] This is why, for the German philosopher Immanuel Kant, for example, the only truly universal science is that of human reason and not of natural philosophy or science.

There is also a version of ahistoricity built into the relational theory of objects. This seems counter-intuitive because relational theorists often take history quite seriously as part of the emergence of objects. However, if virtual relations always precede and exceed their actual *relata*, then the objects themselves are not fully historical. If relations exceed objects, and objects are historical, then relations exceed history. This is precisely why the French relational philosopher Gilles Deleuze explicitly contrasts history with 'pure becoming'.[38]

For the object-oriented ontologist, the withdrawn essence of objects has no history because it is 'cut off' and 'aloof' from all relations with other objects. Since history is about the real movement, change and emergence of objects, and withdrawn essences do not act whatsoever, they are explicitly vacuum-sealed off from history. In this way, all four static theories of objects are also ahistorical.[39]

In response to this, I take a decidedly historical approach. *Theory of the Object* is a history of the material conditions for the emergence of objects. There is not one kind of object or pattern of motion that defines the process of object-creation forever and all time. There can be no single theory of *the* object. Because objects have a history, they also tend to follow different patterns of emergence. This book is not a formal or universal account of objects in general. Nor is it an account of all objects in particular. It is a description of four major *patterns* of historical emergence. New patterns or types may emerge in the future because matter is indeterminate and creative.[40]

What are the consequences?

Three significant contributions follow from my kinetic approach. Each connects to a central part of the book.

Part I: the kinetic object

The first contribution of my kinetic theory of the object is that it offers a few new concepts for describing how objects come into being and change. In the first part of this book I introduce three concepts that explain how objects have been formed so far in history. These three concepts are 'flow', 'fold' and 'field'. In brief, matter flows indeterminately, then folds up and cycles into metastable objects, and is then distributed with others into fields. Objects emerge rather in the way that flows of ocean water fold up into waves, crash, and distribute their patches of bubbles along the shoreline, or like threads weave into fabrics.

I also offer here some new movement-based interpretations of crucial concepts in the history of science such as number, knowledge, reference and observation. These new interpretations help shape my expanded and non-anthropocentric theory of science and epistemology. A movement-oriented theory of science is not grounded in objective forms, subjective experience, withdrawn essences or ontological relations, but instead in the historical and material patterns of movement.

Part II: a history of objects

The second contribution of my theory is that it opens up a whole new way of thinking about the *history* of science and knowledge. It contributes to a process philosophy of science. Unfortunately, as science and technology scholars have been pointing out for some time, most humanities research has tended to eschew serious inquiry into the sciences for at least the last fifty years. Humanists often treat the sciences as if they were the mere reduction of quality to quantity and so they ignore or dismiss them. One of the strengths of new materialism is that it takes the sciences more seriously as unique theoretical practices. This book is part of that effort.

In the search for understanding of how contemporary objects work, I take a historical approach. I want to see how each pattern of motion combines with what came before to create new hybrid objects. There is no truly universal theory of objects because objects are still emerging. The past processes are still immanent and active in objects today and will continue to mutate and mix in the future. When a new kind of object emerges, it does not replace the old but absorbs and redeploys it.

The aim of my history of science is to show that there are four main kinds of objects. The first one I call 'ordinal' because it develops through linear sequences. The second I call 'cardinal' because it creates and organises wholes or units. The third kind of object I call 'intensive' because of its highly differentiated internal structure. The fourth I term 'potential' because of its unspecified or yet-to-be-determined range of possibilities. I cannot show here that mathematics and every single science fit neatly into these four types, but I try my best to bring in as much evidence as I can to show how many of the most important objects do.

The other main feature of my history in Part II is that I connect these four kinds of objects with four distinct kinetic patterns that describe how each type of object was shaped and sustained. For example, for ordinal objects to emerge, I show that people and things had to move in a 'centripetal' pattern, from a peripheral to a central region. A distinctly 'centrifugal' pattern of motion, from the centre to periphery, was required for cardinal objects to emerge. The intensive object formed through relatively rigid 'tensional' kinds of movement and potential objects through relatively 'elastic' kinds of motion. This is a unique way of cutting up history. But if we prioritise motion, I think it follows that non-human objects are much more active than previously thought. Their patterns of motion play a role in their creation and reproduction. *Theory of the Object* is the first study of the agency of objects *as patterns of motion* in the history of science.

New patterns emerge from old ones at different periods in history, but all the older ones persist and mix with the new ones. In the twenty-first century, our objects are hybrids assembled from the building blocks of history. This is why we need a historical approach to understand contemporary objects.

Part III: the contemporary object

The third contribution of my kinetic theory is that it provides a glimpse into the beginnings of a new kind of object found in several contemporary sciences. History is neither linear nor progressive but has coexisting and overlapping patterns of object formation. The contemporary object is an attempt to grapple with this fact. All the ordinal, cardinal, intensive and potential kinds of objects from history are still at work today. Quantum mechanics has not replaced classical mechanics but instead absorbs and transforms it.

I call this new kind of object the 'loop object'. This object's main features are its hybridity, indeterminacy and relationality. The existence of such an object is hardly agreed upon today. The theoretical sciences' cutting edges always have competing interpretations and speculations that drive experimental research. What I offer in this final section is a philosophical synthesis and kinetic interpretation of three contemporary sciences: quantum theory, category theory and chaos theory. My aim is not to challenge these scientific practices or make any new predictive claims, but rather to offer a novel interpretation of what they are doing that shows their common foundation in indeterminate movement.

*

In the next three chapters, we begin our journey into the underworld of objects by thinking about their basic features as material processes. Once this theoretical framework is in place, we can then use it as a new lens to view the history of scientific objects and, finally, understand the most contemporary objects.

Notes

1. Lee Smolin, *The Trouble with Physics: The Rise of String Theory, the Fall of a Science, and What Comes Next* (London: Penguin, 2008), 151: 'From Aristotle up until that point, the universe had always been thought to be static. It might have been created by God, but if so, it hadn't changed since. Einstein was the most creative and successful theoretical physicist of the preceding two centuries, but even he could not imagine the universe as anything but eternal and immutable. We are tempted to say that if Einstein had been a real genius, he might have believed his theory more than his prejudice and predicted the expansion of the universe. But a more productive lesson is just how hard it is for even the most adventurous thinkers to give up beliefs that have been held for millennia.'
2. Before 1929 most physicists, including Albert Einstein, believed that the universe was an immobile spatio-temporal sphere. However, in 1929 Edwin Hubble produced experimental evidence that showed that the universe was expanding in every direction. By 1998 further evidence showed that the universe was expanding at an *accelerating rate*. This fact also changes how we understand motion itself from being extensive (motion from A to B) to an intensive transformation of the whole process. For more on quantum gravity, see Chapter 16.
3. Not all physicists believe that space or time are emergent properties, but many do. There are many quantum gravity theories but none have been experimentally confirmed.
4. Sean Carroll, *Something Deeply Hidden: Quantum Worlds and the Emergence of Spacetime* (Boston: Dutton, 2019), ch. 1.
5. See Chapter 16 on 'quantum entanglement'.
6. Carroll, *Something Deeply Hidden*, ch. 12. See also Karen Barad, 'Transmaterialities: Trans★/matter/realities and Queer Political Imaginings', *Glq: a Journal of Lesbian and Gay Studies*, 21 (2015), 387–422.
7. A.B, 'Using Maths to Explain the Universe', *The Economist*, 2 July 2013, available at <https://www.economist.com/prospero/2013/07/02/using-maths-to-explain-the-universe> (last accessed 4 March 2021).

8. Historical analysis, like science, is not a linear process. Nor is it simply revisionist. It is a feedback loop. Once new historical events occur, then these events provide a new perspective from which to recover a dimension of the past, previously invisible or marginalised. See the concept of 'retroactive certification' in Bruno Latour, 'A Textbook Case Revisited–Knowledge as a Mode of Existence', in Edward Hackett (ed.), *The Handbook of Science and Technology Studies* (Cambridge, MA: MIT Press, 2008), 83–112. See also Thomas Nail, *Being and Motion* (Oxford: Oxford University Press, 2018), chs 1–3 on historical methodology.

9. For a wider treatment of cosmic and earth history, see Thomas Nail, *Theory of the Earth* (Stanford: Stanford University Press, 2021).

10. Thomas Nail, *The Figure of the Migrant* (Stanford: Stanford University Press, 2015); Thomas Nail, *Theory of the Border* (Oxford: Oxford University Press, 2016); Thomas Nail, *Being and Motion* (Oxford: Oxford University Press, 2018); Thomas Nail, *Theory of the Image* (Oxford: Oxford University Press, 2019); Thomas Nail, *Theory of the Earth* (Stanford: Stanford University Press, 2021).

11. Karen Barad, *Meeting the Universe Halfway: Quantum Physics and the Entanglement of Matter and Meaning* (Durham, NC: Duke University Press, 2007); Bruno Latour, *Reassembling the Social: An Introduction to Actor-Network-Theory* (Oxford: Oxford University Press, 2008); Donna Haraway, *Staying with the Trouble: Making Kin in the Chthulucene* (Durham, NC: Duke University Press, 2016).

12. In other books such as the ones on Lucretius and Marx, and others to come, I have chosen depth instead.

13. 'Philosophy [i.e. natural philosophy] is written in this grand book – I mean the Universe – which stands continually open to our gaze, but it cannot be understood unless one first learns to comprehend the language and interpret the characters in which it is written. It is written in the language of mathematics, and its characters are triangles, circles, and other geometrical figures, without which it is humanly impossible to understand a single word of it; without these, one is wandering around in a dark labyrinth.' Galileo Galilei, *Discoveries and Opinions of Galileo*, trans. Stillman Drake (New York: Doubleday, 1957), 237–8.

14. I have in mind here especially Jane Bennett, *Vibrant Matter* (Durham, NC: Duke University Press, 2010), and Thomas Lemke's critique of her metaphysics of relations in 'An Alternative Model of Politics? Prospects and Problems of Jane Bennett's Vital Materialism', *Theory, Culture & Society*, 35.6 (2018), 31–54, doi:10.1177/0263276418757316: 'To put it in an old-fashioned vocabulary: Bennett endorses an "idealist" account of materialism'; 'To put it bluntly: there is a lack of materiality in this vital materialism.' But see also Manuel DeLanda, *Assemblage Theory* (Edinburgh: Edinburgh University Press, 2016) and Latour, *Reassembling the Social*. This is not the place to engage an entire literature review and critique of various relational positions, since I have already treated them in *Being and Motion*, ch. 3, and at length in Christopher

N. Gamble, Joshua S. Hanan and Thomas Nail, 'What is New Materialism?', *Angelaki*, 24.6 (2019), 111–34.

15. Here I also have in mind the work of other process philosophers such as Henri Bergson, Gilles Deleuze and Alfred North Whitehead, whose work is of great interest and inspiration to me and whose theories are perhaps closest to my own. However, my own 'kinetic process philosophy' diverges from each of them on a number of important central points whose full explanation requires its own careful chapter-length treatment and review, which would be redundant to reproduce here since it is already published as chapter 3 of Nail, *Being and Motion*.

16. For a critique of the idea of change without motion, see Nail, *Being and Motion*, ch. 3.

17. Bruno Latour and Albena Yaneva, '"Give Me a Gun and I Will Make All Buildings Move": An Ant's View of Architecture', in R. Gesier (ed.), *Explorations in Architecture: Teaching, Design, Research* (Basel: Birkhäuser Verlag, 2008), 80–9 (81).

18. Graham Harman, Bruno Latour and Peter Erdely, *Prince and the Wolf: Latour and Harman at the LSE* (Lanham, MD: Zero Books, 2011), 44: 'And you say that if one is an occasionalist (and since you now give me this beautiful name I will accept it) then one has to be an occasionalist all the way.'

19. Graham Harman, 'Buildings Are Not Processes: A Disagreement with Latour and Yaneva', *Ardeth*, 1.9 (2017), 112–22 (117).

20. Graham Harman, *Immaterialism* (Cambridge: Polity, 2016), 13, 15.

21. Object-oriented ontology (OOO), a term coined by Graham Harman, defines a theoretical commitment to thinking the real beyond human experience. As such, the reality of matter is never something anthropocentric, experienced or relational, but always something that 'withdraws'. This leads Harman, like Badiou, to affirm what they call 'a new sort of "formalism"'. Timothy Morton similarly argues against 'some kind of substrate, or some kind of unformed matter' in favour of infinitely withdrawn essential forms. See Thomas Lemke, 'Materialism Without Matter: the Recurrence of Subjectivism in Object-Oriented Ontology', *Distinktion*, 18.2 (2017), 133–52. See also Carol A. Taylor, 'Close Encounters of a Critical Kind: A Diffractive Musing in Between New Material Feminism and Object-Oriented Ontology', *Cultural Studies*, 16.2 (2016), 201–12.

22. Harman, *Immaterialism*, 16.

23. Harman, *Immaterialism*, 27–9.

24. Graham Harman, 'On Vicarious Causation', *Collapse*, II (2012), 211.

25. Inside are 'explosive undercurrents belonging only to individual things, withdrawn from full expression in the world'. Graham Harman, 'Agential and Speculative Realism: Remarks on Barad's Ontology', *Rhizomes*, 30 (2016), available at <http://www.rhizomes.net/issue30/harman.html> (last accessed 4 March 2021).

26. Graham Harman, *Prince of Networks: Bruno Latour and Metaphysics* (Melbourne: re.press, 2009), 187.

27. Harman, *Immaterialism*, 16.
28. Graham Harman, 'Materialism is Not the Solution: On Matter, Form, and Mimesis', *The Nordic Journal of Aesthetics*, 47 (2014), 94–110 (100).
29. Harman, *Immaterialism*, 47.
30. 'Everything has an autonomous essence, however transient it may be.' Harman, *Immaterialism*, 16.
31. Harman, 'Agential and Speculative Realism'.
32. Harman says 'I would like to volunteer OOO to serve as a model of what they [Bruno Latour and Albena Yaneva] and call static architecture.' Harman. 'Buildings Are Not Processes', 116.
33. Latour and Harman are 'secular occasionalists' who believe that change occurs discontinuously without material movement. For Latour it is the networks that change discontinuously, and for Harman it is the objects that change discontinuously. 'In this way, Bruno Latour is the first *secular occasionalist*: the founder of what I have called vicarious causation.' Harman, *Prince of Networks*, 115. Harman also argues that the French philosopher and founder of speculative realism Quentin Meillassoux is 'the most extreme occasionalist who has ever lived'. Graham Harman, *Quentin Meillassoux: Philosophy in the Making* (Edinburgh: Edinburgh University Press, 2011), 144.

    I am not the only one to describe Harman's theory of change as static. Steven Shaviro says, 'Harman accounts for change by appealing to the emergence of qualities that were previously submerged in the depths of objects; but he does not explain how those objects came to be, or how their hidden properties got there in the first place.' Steven Shaviro, 'The Actual Volcano: Whitehead, Harman, and the Problem of Relations', in Levi R. Bryant, Nick Srnicek and Graham Harman (eds), *The Speculative Turn: Continental Materialism and Realism* (Melbourne: re.press, 2011), 285. Shaviro continues, 'Harman's entities, in contrast, do not spontaneously act or decide; they simply are. For Harman, the qualities of an entity somehow already pre-exist; for Whitehead, these qualities are generated on the fly. Harman, as we have seen, discounts relations as inessential; his ontology is too static to make sense of them' (287).

    For a critique of OOO's theory of change, see also C. J. Davies, 'The Problem of Causality in Object-Oriented Ontology', *Open Philosophy*, 2.1 (2019), 98–107; Lemke, 'Materialism Without Matter'. I thank Christopher N. Gamble for talking through OOO's static ontology with me.
34. See Carroll, *Something Deeply Hidden*.
35. Harman, 'Agential and Speculative Realism'.
36. For a more detailed account of the difference between my process philosophy and Bergson's, Whitehead's and Deleuze's, see Nail, *Being and Motion*, ch. 3.
37. While they remain anthropocentric, they also allow for radical historical changes in existing social and aesthetic structures, unlike strict Kantian subjective idealists. See Walter Benjamin, 'Theses on the Philosophy of History', in *Illuminations: Essays and Reflections* (New York: Schocken Books, 1969), 253–64. See also Theodor Adorno, *History and Freedom: Lectures 1964–1965*, trans. Rolf Tiedemann (New York: John Wiley and Sons, 2014).

38. See the historical critique of Deleuze in Thomas Nail, 'The Ontology of Motion', *Qui Parle*, 27.1 (2018), 47–76.
39. To my knowledge there are histories of science, and histories of philosophy, but not histories of 'the philosophy of objects' that take seriously the historicity of objectification as a real material and historical process. Marxist histories get the closest, but privilege the social over the objective. See Nail, *Theory of the Image*, and Thomas Nail, *Marx in Motion: A New Materialist Marxism* (Oxford: Oxford University Press, 2020).
40. Again, the complete methodology at work here can be found in Nail, *Being and Motion*.

PART I
THE KINETIC OBJECT

1

The Flow of Matter

What is the kinetic theory of the object? It is a way of interpreting or understanding objects as processes of movement. We may find it useful sometimes to treat objects as if they were static, but when we do we tend to overlook what creates, sustains and changes them. We tend to treat objects as products and not processes. However, overlooking this can lead us to some of the critical errors in theory and practice that I discussed in the introduction to this book.[1] This chapter and the next two aim to reverse this view.

Every product presupposes a process. All objects come from somewhere and dissolve into something else. They persist only momentarily in various forms because some activity is continually running through them. Part I of this book attempts to define three general features of the processes that produce objects, which seem to apply to all the objects we are aware of so far in history. I remind the reader again that this is not a universal theory. It is the product of the specific historical study I undertook in this book.

Together, I call these three concepts of object-production a 'kinemetrics' because they define objects as acts or processes of 'measurement-in-motion'. This is a different perspective on objects than usual. It also offers a more material and performative way to think about what the quantitative sciences are up to when they measure and order objects.

The next three chapters aim to give a very synthetic account of the three fundamental processes that create, sustain and order objects. The first of these processes I call 'flow'.

What is flow?

Objects flow. They are thrown into the world, sculpted into forms, and dissolved again into other objects. But what is the source of this movement and flow?

Objects are matter thrown into motion. Without the constant flow of energy through things, there is no accumulation of matter into stable objects. The transfer of energy into and out of quasi-equilibrium states is what allows objects to emerge and persist. It is also what distinguishes them from one another.

This is the case at every scale of material reality. Matter must flow, or else objects would never emerge or dissipate. At the quantum level, fields vibrate into metastable particles. At the classical level, atoms vibrate in patterns we call molecules, and so on up to the cosmos. As far as we know, there is nothing in the universe that does not flow from high concentrations of energy into lower ones. The universe tends to spread out from hot to cold, and this is the source of all flow, movement and process. All objects flow, therefore, because they are part of the moving, material universe.

However, what I call the 'flow' of matter that traverses all objects is *not an object*. Vacuum fluctuations in quantum physics, for example, are not objects. They cannot be measured or observed directly or determinately, and therefore are not objects, in my view. However, this does not mean that they are a lump of unformed matter either. A flow is a process, and as a process is not a determinate 'something'. A flow is neither formed nor unformed but is rather indeterminate. If all material objects are made of indeterminate quantum processes, then all objects are never fully reducible to something static. Their identity is always in flux to some degree, no matter how trivial. Even macroscopic processes such as air currents are still 'flows', because they are in motion at some level. There is no inert substance called matter because nothing is ever entirely still or static, nor reducible to space or time at the quantum level.[2]

Objects, however, did not come from flows like chickens from eggs. There was not first some pure primordial matter without any objects from which the cosmos emerged. Even the densest high-energy state we can think of, the big bang, probably had a finite size that we could theoretically measure in 'Planck units'.[3] A Planck unit is the smallest physical object that can be mathematically measured. Even in the big bang, flows and objects coexisted as two aspects of the 'same' indeterminate process.

Objects come into being through material processes that are immanent to them, as whirlpools emerge from water going down a drain. Objects are not just provisional mental constructs that happen only in the minds of human scientists. In the broadest sense, objects are something that matter makes with or without humans. A river is an object produced by the accumulation of rain flowing into a basin of attraction. The more the water accumulates in the basin, the more it deepens and widens the basin, and the more water tends to gather, and so on, in a feedback loop. The river is a process-object created without human intervention.

However, humans can accomplish similar material accumulations by marking tallies on bones. Tallies can measure a 'lunar cycle' in an object.

Even gathering sticks into a pile and lighting them can create the process-object, 'fire'. In this way, flows of matter accumulate and stabilise into relatively distinct objects. Flows do not precede objects nor do objects precede flows, but both emerge and change together in the 'process of objectification'.

Objects only emerge because matter can encounter itself, or act on itself. Matter is only able to affect itself if it differentiates itself in some way from itself by *flowing*. In other words, matter only becomes an object when a flow can return to itself in a relatively periodic or differentiated pattern. If not, it would immediately dissipate, following the law of entropy. Therefore, all objects must be in continual motion at some level, or nothing, by definition, would be able to exist or persist.

What is matter?

Matter is a process, not a substance. As a process or 'flow', it is neither one thing nor another but is indeterminate.[4] It is neither wholly unformed nor determinately formed, but is continually transitioning *between forms*. Objects are material, but also emerge, persist and dissipate within this meta-morphic matter like waves on the ocean.

This is not a universal definition but a historical one. 'Matter' is our historical name for what is in motion. The tricky part though is that if matter *is in process*, it is also indeterminate. This makes it impossible to give a strictly empirical or metaphysical definition of what matter 'is' because matter has no fixed being. It is always 'matter-in-motion'.

But what is motion if it is not the *determinate* motion of an object from point A to point B? If there are no fixed points such as A or B nor a self-identical object that moves, how can we speak of movement? Here it is helpful to distinguish between two types of movement, both of which happen simultaneously. The first we can call an 'extensive' movement. This is where a relatively stable object moves from one relatively stable point A to another relatively stable point B. For example, we can watch a butterfly flutter from one flower to another. However, there is also another kind of movement we can call 'intensive'. This other type of movement is when the whole AB transforms, as seasons change, or as a pebble thrown into a puddle transforms the entire pool.

Let's look at an example from physics to illustrate how both of these kinds of motion are active simultaneously. If we look at an atom under a powerful microscope, we can map its trajectory as it moves from one point to another. This is its *extensive* motion. However, inside this atom there are also electrons, and these electrons are made of indeterminately vibrating fields of energy. Our microscope, our bodies and all of nature are also made of these same fields and are changing at the same time. So what appears to be a relatively isolated self-identical particle moving from A to B in 'empty space' is also an entangled transformation of many quantum fields at once. When all the fields change together, this is an *intensive* motion. In this way, motion is both extensive and intensive at the same time. They are two aspects or dimensions of all motion.

This is also why matter is not entirely empirical or observable. We cannot directly measure or observe the quantum fluctuations of objects. We see their results and changes, but only obliquely as visible changes in objects. Just as the steamy vortex you see spiralling from your hot coffee is an object that is inseparable from the flowing moisture that forms it, so non-objective flows of matter are inseparable from their objects. There is no ultimate stuff that we can observe and that defines the essence of 'pure matter', only vibrating indeterminate flows.[5]

A material flow, in my view, is anything with energy and momentum. Quantum fields, for example, are not observable objects because their energy and momentum are indeterminate.[6] Unlike classical matter, fields of matter are not entirely measurable or predictable. Quantum fields are creative, unstable and in constant motion.[7] They do not follow causal laws but rather move stochastically in irregular but patterned frequencies. If all matter has this quantum feature, then at some level, and to some degree, all matter 'flows'.[8] And if objects are material then they all flow to some degree.

Finally, a flow of matter is neither continuous nor discrete. This is because matter is not a substance but an indeterminate process. Matter, space and time are emergent features of reality that occur in and through flows.

What is pedesis?

Pedesis is an unpredictable movement. All material movement begins with indeterminate quantum fluctuations, but even at macroscopic levels, these fluctuations can give rise to non-trivial changes that ripple through whole fields of objects.[9] All movement is pedetic to some degree. This is the reason that matter does not flow in a straight line or follow iron-clad deterministic laws. It 'swerves', as the Roman poet Lucretius stated two thousand years ago.[10] The unpredictable swerve of matter allows it to converge and diverge in new and creative ways. The swerve is also what allows matter to engage with itself in the metastable patterns that we call 'objects'.[11]

Our modern understanding of pedesis comes from two critical discoveries in twentieth-century physics: Einstein's kinetic theory of matter (1905) and Heisenberg's quantum uncertainty principle (1927). Einstein was the first to experimentally show that the classical objects of daily life, which

appear so solid and stable, are, in reality, composed of much smaller pedetically moving objects such as molecules and atoms. Gases, for example, Einstein argued, are objects whose component objects are moving around faster
and farther than those of fluids, and fluids are objects whose components
are moving faster and farther than those of solids. Einstein showed that all
objects were metastable patterns made from pedetic motions. He established
that the forms of things were emergent features of matter-in-motion.

This discovery introduced turbulence into the heart of all the objects
we once took to be stable and predictable. Although the first account of
turbulence also goes back to Lucretius' description of dust motes swirling
in the sunlight, physicists still do not have a successful predictive theory of
turbulent motion. Turbulence's precise kinetic structure thus remains one
of the last unsolved problems of classical physics, with a million–dollar prize
for its mathematical solution.[12]

The unsolved problem of turbulence, combined with Einstein's kinetic
theory of matter, has a radical consequence. All matter is in motion, and
all motion has turbulence with no deterministic solution. Heisenberg once
said he wanted to ask God two questions. The first was, 'Why is general
relativity so weird?' The second was, 'How do you explain turbulence?'
He then said that he was certain God would know the answer to the first
question.[13]

The second great kinetic discovery of the twentieth century was when
Heisenberg showed a fundamental limit to the precision with which the
position and momentum of a particle, such as an electron, can be known at
the same time. The more precisely the position of a quantum field is determined, the more it looks like a stable particle, but the more its momentum
becomes uncertain. Conversely, the more we determine the momentum,
the more its position becomes uncertain. In other words, the precise extensive path of a particle from A to B is *fundamentally* uncertain.[14]

This does not mean, however, that motion is necessarily random. Position and momentum occur only in *relationship and response* to the whole
surrounding quantum apparatus. What is interesting to me about the
movement of matter here is not merely that it is pedetic, but that through
pedesis and turbulence, metastable formations and objects emerge. If all
movement is random, then how could objects hold together for more
than a second? For movement to be wholly and genuinely random, matter
would also have to be unaffected by anything that would hold it in place.[15]
This would seem to violate the laws of physics as we currently know them.
Furthermore, given the high level of order and complexity we see around
us, this kind of pure randomness does not seem likely.

Pedetic motion, on the other hand, is not random. It emerges from
other motions, although not in a wholly deterministic way. Pedetic motion

and random motion are both unpredictable, but for different reasons. Random motion is unpredictable because no other motions influence it. Pedetic motion, however, is fundamentally unpredictable *because so many other motions affect it*. In pedesis, it is the interrelation, intra-action and mutual influence of matter with itself that makes it unpredictable. Since pedetic movements are responsive and relational, they can also combine and stabilise into patterns without randomly breaking apart. These patterns are what we call objects. They are metastable structures with relative degrees of stability and solidity. However, pedesis also means that the flows of matter are inherently unstable and tend to spread out over time, entropically. Objects, therefore, tend to dissipate.

Turbulence is a beautiful image of how objects emerge, persist and dissipate unpredictably. Turbulent fluids appear chaotic, but they are not random. Over time and under specific conditions, patterns such as vortices and branching shapes develop, persist and eventually dissipate. Objects come into being and pass away like this.

Matter is pedetic not only at the quantum level of indeterminacy but also at macroscopic levels as well. We find turbulence and vortices from quantum fluids and laser light to star and galaxy formation. Quantum and classical vortices differ in some ways, but they also paint a common picture of how metastable forms tend to emerge from processes. Pedetic movements can crystallise into ordered patterns at every level of nature, even if they take millions of years to appear. Pedesis always generates some bit of order.[16] That order is the object.

Conclusion

I have tried to show in this chapter that objects emerge and persist through flows of matter. I defined a flow as an indeterminate process of energy and momentum that is not reducible to any directly observable object. However, through their pedesis and turbulence, flows also fold and unfold into metastable patterns we call 'objects'. As such, flows always coexist with objects just as the river is inseparable from its whirls and eddies.

'Flow' is the first concept of my kinetic theory, and in the next chapter I would like to show more precisely how these pedetic flows of matter can fold into relatively stable objects.

Notes

1. For a description of some of these errors, see Thomas Nail, *Theory of the Earth* (Stanford: Stanford University Press, 2021), and Thomas Nail, *Lucretius II: An Ethics of Motion* (Edinburgh: Edinburgh University Press, 2020).

2. See Karen Barad, *Meeting the Universe Halfway: Quantum Physics and the Entanglement of Matter and Meaning* (Durham, NC: Duke University Press, 2007), 422, n. 15. Carlo Rovelli, *Reality Is Not What It Seems: The Elementary Structure of Things*, trans. Simon Carnell and Erica Segre (New York: Penguin, 2017), 132–3: 'An electron, a quantum of a field, a photon, does not follow a trajectory in space but appears in a given place and at a given time when colliding with something else. When and where will they appear? There is no way of knowing with certainty. Quantum mechanics introduces an elementary indeterminacy to the heart of the world. The future is genuinely unpredictable . . . The world is not made up of tiny pebbles. It is a world of vibrations, a continuous fluctuation, a microscopic swarming of fleeting microevents.'

3. See Rovelli, *Reality Is Not What It Seems*, 207–9.

4. Following the indeterministic interpretation of quantum physics. See Chapter 16.

5. Again, this is only the case given the ontological primacy of motion in the present. On the question of whether motion and matter will always be primary or whether something else will emerge, I remain agnostic. Any such speculation is necessarily metaphysical.

6. 'The way to see the energy/momentum of a field is to arrange some clever experiment in which a series of "microscopic" movements of energy and momentum in the field kick off a chain reaction of larger-scale movements of energy/momentum until a "macroscopic" thing is affected in a way that we can see. This is basically what designing an experiment is all about.' Personal correspondence with Brian Skinner, a researcher in theoretical condensed matter physics at MIT.

7. See Sean Carroll, *The Particle at the End of the Universe: How the Hunt for the Higgs Boson Leads Us to the Edge of a New World* (New York: Dutton, 2012).

8. Quantum fields can only be observed through the visible effects they create and not in themselves. In order to generate mass and particles, quantum fields by necessity must have energy and momentum. Since, as Einstein showed, mass and energy are convertible, particles are born from and return to their quantum fields. Quantum field energy becomes particle mass, becomes field energy, in a continuous momentum or movement. Therefore a quantum field is just as material as particles are – even if the field itself is not empirically visible – because *particles are nothing other than folds or excitations in flow of fields*. Matter therefore is always already a flow of matter that has simply folded up into a particle.

9. See Chapter 18.

10. See Thomas Nail, *Lucretius I: An Ontology of Motion* (Edinburgh: Edinburgh University Press, 2018); Thomas Nail, *Lucretius II: An Ethics of Motion* (Edinburgh: Edinburgh University Press, 2020); Thomas Nail, *Lucretius III: A History of Motion* (Edinburgh: Edinburgh University Press, forthcoming 2022).

11. Like the double spiral vortex of Taylor and Couette's double cylinder flow in fluid dynamics. G. I. Taylor, 'Stability of a Viscous Liquid Contained between

Two Rotating Cylinders', *Phil. Trans. Royal Society*, A223 (605–15) (1923), 289–343, doi:10.1098/rsta.1923.0008

12. See the Clay Mathematics Institute (CMI), available at <http://www.claymath.org/millennium-problems/navier–stokes-equation> (last accessed 4 March 2021).

13. The quote is probably apocryphal, but the sentiment is striking.

14. For a non-epistemological interpretation of uncertainty and indeterminacy, see Rovelli, *Reality Is Not What It Seems*, 132–4. See Barad, *Meeting the Universe Halfway*, 301, for an interesting interpretation of uncertainty beyond Bohr's idea of limiting observation to the walls of the laboratory.

15. The 'Boltzmann brain' is named after the nineteenth-century physicist Ludwig Boltzmann.

16. Consider also the likelihood of certain recurring patterns in the case of a cosmic big bounce scenario put forward by Carlo Rovelli and others, or in the case of the multiverse put forward by Sean Carroll and others.

2

The Fold of Number

Objects are flows of matter that have 'folded' into relatively stable patterns or cycles, such as vortices. As a flow cycles around and around an area, we can measure its relative stability, size and speed in space and time. How far does it go out before coming back? How fast does it loop around? How many times does it loop around? Objects, in my view, are not static quantities, forms or essences, but rather what we might call 'metastable quantities'. As such, we have to measure and understand them quite differently than fixed objects. If objects are process, what does that mean for our understanding of measurement, quantity and number more generally? This is the subject of this chapter.

If the objects we see around us are not self-identical unities, but more like loops, why do they look so unified and stable? Each time a flow of matter goes out and returns, I call this its 'loop' or 'fold'. Objects are steady oscillations or folds of matter like dynamic equilibriums or stabilities-in-motion. The looping and folding dimension of objects is their 'quantity' because it can be coordinated with other cycles and 'counted'. But what does counting mean here without a fixed unit or object to count? This chapter answers this question with a process theory of number.

In the previous chapter, I focused on describing the material processes or 'flows' that comprise objects. This chapter describes how such indeterminate processes can become quantitatively distinct from one another and conjoin to produce larger composite objects.

What is number?

By the term 'number', I mean the quantitative dimension of objects. It is the 'how much' or 'magnitude' of things measured in space and time. I realise that this is a much broader definition of number than we are used to, but I think it helps us overcome a couple of problems I will discuss shortly. As I describe it here, number is not just natural numbers such as 1, 2 or 3, but is the basic structure of magnitude more generally. There are no numbers, arithmetic or even abstract numbers such as the square root of 2 or imaginary numbers without this more basic magnitude.[1] All numbers assume a more fundamental quantitative dimension that I want to rethink from a more kinetic perspective.

In my material kinetic view, quantity is not just an abstract idea humans came up with to count things. It is a genuinely emergent dimension of reality. All kinds of things oscillate back and forth to varying degrees. The path or shape of these oscillations or loops is what distinguishes them from one another. For example, a flower can produce five distinct petals of a particular length and width different from other flowers with six petals. The molecules in these petals also vibrate back and forth with a specific frequency that sustains their degree of solidity and their unique extension in space and time. This idea is the gist of Einstein's kinetic theory of matter, as discussed in the last chapter. In short, as matter moves, it shapes the space and time of objects.[2]

The quantity of an object thus depends on how a flow of matter cycles or iterates. However, most definitions of number tend to treat it as a mental symbol that humans invented for keeping track of things. But where did humans get such an idea in the first place? I think if we start to look more carefully at the material conditions that made this human idea possible, it is much harder to say that number is *just* an idea. This is the project of Part II.

From a materialist perspective, it is not enough to say, as the Irish philosopher George Berkeley once did, that number is 'nothing but names'.[3] Nor is it sufficient to claim that number is merely an idea in human minds as the German mathematician Georg Cantor claimed.[4] I do not think that numbers are just useful conventions, and I certainly do not believe they are essences that reside in an unchanging realm beyond the material world.

All of these positions I have just listed, 'nominalism', 'mentalism', 'fictionalism' and 'Platonism', only raise the deeper question of how, and under what conditions, humans came to define number in these ways. The history of our theories of number presupposes the very thing it pretends to explain: how number emerged. In the non-Platonic positions, number is something humans invented, but this only pushes the problem further back. How were humans able to count at all? For the Platonists, the ideal existence of number is merely asserted, like the self-evident existence of ideal forms more generally. Their position is that number is an ideal form, and ideal forms have immaterial existence; therefore, number has immaterial existence.

However, in my view, at the foundation of these positions lies a more material and historical description that accounts for both the kinetic basis of number and the emergence of various subjectivist and metaphysical philosophies of number. This chapter on the fold aims to provide a material kinetic theory of number as a core feature of objects. The full historical basis of this theory will be historically developed further in Part II.

First of all, we must at least state the obvious: humans are not the only beings that can count. Several recent studies have shown a wide range of animals and insects that can count and even do arithmetic.[5] They might not use cardinal numbers such as 1, 2 and 3. Still, they are nevertheless fully capable of distinguishing discrete objects and changes in objects' magnitude, something foundational for all higher mathematics.

Without the capacity to *sense* magnitude, there is no number. The solitary female wasp, for example, lays her eggs in individual cells and feeds each egg with a specific number of live caterpillars on which the young feed when hatched. Some species place five, others twelve, and others twenty-four caterpillars in each cell. In the genus *Eumenes*, the mother wasp even distinguishes between male and female eggs, giving different numbers to each sex.[6] All mammals, and most birds, will notice a change in the number of their young nearby. Many birds can distinguish two from three and will abandon a nest when a certain number of eggs are missing, but not before.[7]

The foundations of number thus lie in *number sense*: the ability to recognise differences in magnitude.[8] All objects have a quantity or magnitude measured relative to other magnitudes, but if there are no absolute divisions in matter, what distinguishes one object from another? The answer is that all objects have different qualities depending on how matter cycles or folds through them, as described above. There is no magnitude without qualitative sensation, and thus no difference in magnitude, and hence no number. As Hegel says,

> The truly philosophical science of mathematics as *theory of magnitude*, would be the science of *measures*; but this already presupposes the real particularity of things, which is found only in concrete Nature. On account of the *external* nature of magnitude, this would certainly also be the most difficult of all sciences.[9]

The theory of magnitude, or what I am calling 'number', is the science of measurement. But what is measurement? It is the rhythmic coordination of distinct processes. As such, it requires, first of all, the existence of qualitatively different objects. In other words, before there can be a human, animal or insect sensation of a difference in magnitude, nature must already be something with real qualitative and quantitative differences. It must already be self-affective or 'folded', in my terminology. Qualitative magnitude must be something nature already has, in a certain sense, or else insects, animals and humans could not coordinate different qualities to one another. In other words, matter must be able to make and order objects to some degree.

As Hegel says, the counting of magnitude 'already presupposes the real particularity of things'. This includes the difference between what counts and what is counted. Even the process itself of counting must be a distinct kinetic act.[10]

This materialist definition of 'number' and 'counting' includes a much more comprehensive range of objects than typically thought. In this definition, we can consider all kinds of oscillatory processes as 'counting' or 'measuring'. For example, pacemaker, circadian and clock cells in organisms transmit chemical signals in coupled patterns with other cells.[11] These cells establish temporal measurements such as heartbeats and sleep–waking durations and measure out how big a body can grow. In this way and others, living objects are distinct from one another because of internal systems of periodic oscillation. These include the synchronised flow of blood and neuronal activity and the rate of cellular regeneration.

Furthermore, biological networks of coupled oscillations are also composed of oscillating molecules vibrating back and forth in different periodic orbits. The frequencies of these orbits distinguish molecular objects kinetically from others and give rise to biological cells' chemical unity. In turn, these molecules are composed of vibrating atoms with oscillating orbital electrons, and so on. All of nature is composed of vibrating and oscillating matter. Even minerals differentiate themselves through the coupled oscillatory speed of their molecules. Phase transitions between solids, liquids and gases are material changes in the measure, number and quantity of objects. These iterating frequencies are what I call the 'folds' of objects.

Oscillation 'keeps count' by vibrating back and forth, going out and returning at specific, although not always identical, intervals or cycles. These intervals produce a magnitude of duration and extension. In this sense, the flows of matter 'count' themselves each time they oscillate. Counting, in this expanded kinetic definition, is, therefore, not an exclusively mental activity. Thousands of clock cells occur in biological systems. Even pendulum clocks hung on the same wall will start to sync with one another. We can thus say with Hegel that nature is 'pure externalisation' and 'magnitude', but only on the condition that we say with Marx that this externalisation emerges only *through motion*.[12]

If all objects consist of matter flows, then folds are the kinetic structures of relative stasis or stability. A fold is like a spiralling storm system. It is a redirection of a flow of matter back on to itself in a loop. The fold is the process of self-affection or 'self-numeration'. The fold of the object is not merely the product of a flow because there is never any final product. The fold is always folding-in-motion.

Folding is repetition with a difference. It is a vortical process that repeats in approximately, but not precisely, the same looping pattern, in mobile stability or homeorhesis.[13]

The region where a flow of matter intersects with itself is what I call the 'period' of the object. Although the flow of matter is continually changing and moving around the loop, its period can remain in the same place, like the centre of a metastable vortex.

The time and space that it takes for a flow of matter to return to itself I call the 'cycle' of the object. The path the flow takes is how the object measures or 'counts' itself. Each cycle is an iterative measure of how long it takes for a flow to return to its starting point and how far it goes before it returns.

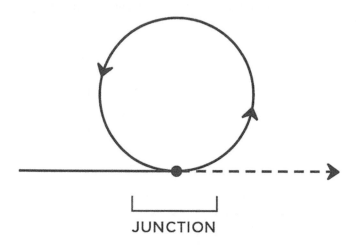

JUNCTION

Figure 2.1 Fold and junction. Author's drawing.

In this kinetic definition, objects are not strictly identical to themselves. Objects have no essences because, as metastable formations, they are slightly different in each iteration. These cycles do not create perfect regularities or equilibrium but work more like 'attractors'. An attractor is a region that movements tend towards with frequency but not strict identity. Therefore, objects are not discrete blocks, but areas where flows of matter tend to overlap over time like a Lorenz attractor.

Objects are only approximately identical at the macroscopic level, but increasingly differential at microscopic levels.

My diagram cannot show it, but the reader can imagine these loops as flows that are continually moving around the loop. The 'period' does not stay in the same place each time but moves around relative to the loop. In

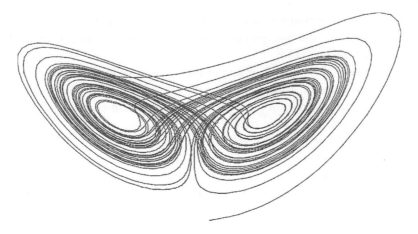

Figure 2.2 Lorenz attractor. https://commons.wikimedia.org/wiki/File:Lorenz_attractor2.svg.

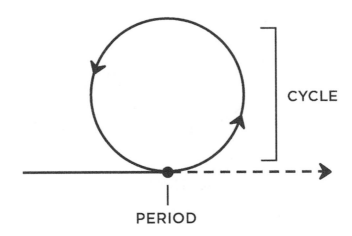

Figure 2.3 Cycle and period. Author's drawing.

this way, objects are fundamentally unstable and resist geometrical abstraction. This is why computer-generated fractals are only rough approximations of unique and relational natural fractals.

Real measurement is much more like the dynamics of a dripping tap. Suppose we count only the number of drips in an hour. In that case, we will fail to see that each drop varies slightly in size and frequency based on many internal and relational dynamics with its milieu. This is what Robert Shaw and Peter Scott demonstrated in 1984 in their study of the irregular periodicity of a dripping tap in their laboratory at the University of California, Santa Cruz.[14]

Even the circadian cells within all humans, animals and plants do not cycle at precisely regular intervals but respond to the sun, temperature and season. Without these influences, human circadian rhythms and body temperatures, for example, will produce all manner of different cycle lengths from 25 to 45 hours.[15] Cycles never repeat precisely and so objects are never strictly identical.

Physicists have recently discovered a similar phenomenon in quantum mechanics.[16] We now have experimental proof that entangled particles can periodically 'return home' to their initial positions instead of dispersing through space. This surprising phenomenon runs counter to the common interpretive assumption that quantum processes are fundamentally random. Instead, particles can create periodic patterns as if scarred or leaving a trail of breadcrumbs behind them. Whence its official name, 'quantum scarring'.

Objects are always in constant movement because they are continually receiving new motion from outside while losing some of that motion as it passes through them. Therefore, objects have no fixed period or cycle, but only more or less dense periodic orbits or 'limit cycles' that continue to shift around, like the almost 200-year-old Great Red Spot of Jupiter or the North Pacific Gyre. So we *can* step in the same river twice, when the river turns over itself in local eddies and whirlpools.[17] The periodic fold remains the *same* but only on the condition that *others* flow through it. As Heraclitus writes, 'On those stepping into rivers staying the same, other and other waters flow.'[18] For Heraclitus, each eddy in the river is like another river within the 'same' river. A cycle of the object is not a static unity but a fluid or 'kinetic unity'. Just like a whirlpool in a river, a cycle is only a metastable unity of a differential process refreshed each time with new water around a periodic attractor.

Numeration

A number can also contain more than one cycle. Each cycle departs from and returns to the same period in larger or smaller intercalated loops. As each cycle returns to the same periodic point, it reproduces the cycle's identity with itself. It reproduces the unity of all the intercalated cycles with the same periodicity. This is what I call 'numeration'.

How many cycles or oscillations does an object have? It is only possible to count or measure them relative to another cyclical object. This is what Einstein's theory of 'general relativity' tells us. As bodies move away from gravitation centres, the material oscillations that compose their bodies *slow down* relative to others closer to the gravitational centre. But each body does not feel itself as going any faster or slower because its oscillations all slow down at the same rate. This is not something that only 'seems' to

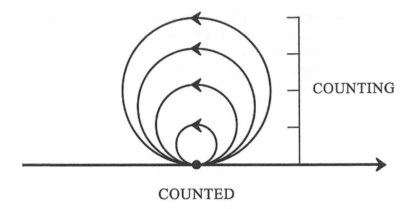

COUNTING

COUNTED

Figure 2.4 Intercalation and numeration. Author's drawing.

be the case because of human psychology. It is the real kinetic relativism of the world. Relative to a body close to the speed of light, Einstein's 'special theory of relativity' tells us that another body on earth *really* lasts only as long as a single heartbeat for the moving body.

In a different, simpler biological example, consider that our fastest arm motion to swat a fly is to the fly an easily avoidable, slow-moving obstacle – relative to the speed of its evasion. When something measures a difference in magnitude, it is always a difference *relative* to the measuring body's own temporal and spatial magnitude.

In short, if all objects are in motion, and all motions are relative to other motions, then all magnitudes are relative to other magnitudes. Magnitude is just a magnitude measuring a magnitude. This does not mean that magnitudes are not real or accurate. They genuinely express a specific and precise kinetic magnitude.

A magnitude is a single but differentiated cycle. What I am calling 'numeration' here is the name for the *kinetic difference* within each cycle of counting. I mean that any given 'count' always has at least three performative dimensions (the counted, the counter and the count), as the second-century Roman mathematician Nicomachus of Gerasa noticed long ago.[19] This means that there never is a real unity or identity of what is counted. There is always a process of multi*plication* or mani*folding*. There is no such thing as a discrete 'one', but invariably a series of material processes of folding and unfolding in counting. Usually, though, we pretend that we are dealing with discrete unities that are real regardless of anything else around.

Magnitude persists only as long as the material oscillations that compose it keep on cycling in approximately the same periodic orbit. As long as the molecules in an ice cube cycle at roughly the same speed, the cube will remain solid. However, as soon as they begin to cycle faster, they will start melting into water.

Numeration is, therefore, something performative and enacted and not just given. It requires a particular kinetic structure relative to other objects. Number is not something that only humans do in their 'mind', whatever that might be. The human brain is only capable of counting at all because of the cyclical and periodic motions of its material neuronal and biological patterns. Brain matter 'counts' itself through persistent firings with various durations and oscillations, all of which precedes any idea we might have 'about' numbers, e.g., 1, 2, 3 . . .[20] But such oscillations are not unique to the human brain. All of nature oscillates.

Qualities and quantities

All objects have qualities and quantities. The two are inseparable dimensions of the same movement of matter.[21] In my kinetic theory, a quality is the 'period' where the flow of matter intersects or affects itself, and quantity is the 'cycle' or path traced by the flow as it completes its loop. The two go hand in hand like art and science. By counting the number of smaller subcycles, we can determine greater and lesser quantities.

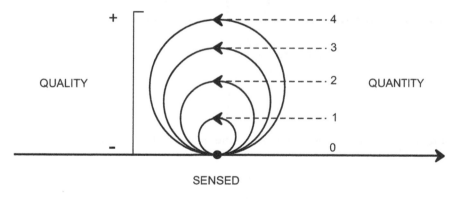

Figure 2.5 Quality and quantity. Author's drawing.

For example, 10 degrees in temperature is hotter than at least nine other measurable qualitative subcycles, or degrees. In this way, we can count a cycle as a quantity without any division between the object's period and cycle. For example, contemporary physics accepts both the qualitative flux of matter as quantum fields and the quantifications of those fields at different emergent levels: particles, atoms, molecules, cells, animals, plants, galaxies and so on.[22] This is possible only because quantity is nothing more than the kinetic cycle of matter-in-motion, considered as a unity or 'one'.

Quantity is like a movement of expansion of the kinetic period to the whole cycle, while quality is like a movement of contraction of the cycle to the single point of its self-sensation or affection. In this way, quantity and quality are the two dimensions of objects.

What is a conjunction?

A conjunction is an assembly of objects. In other words, folds can be looped or woven together into a composite fabric. I call the conjunction of folded objects a 'thing'. Just as individual folds persist as patterns, woven assemblies of these folds can also endure as metastable composites or 'things'. Things conjoin qualitative and quantitative aspects. When we are talking about just the qualitative features of things, we call these 'images'.[23] When we want to refer to only the quantitative aspect of things, we call them 'objects'.

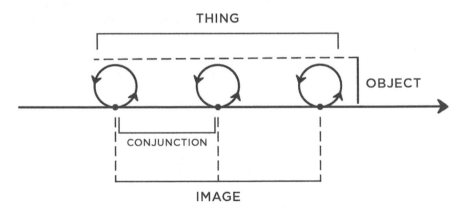

Figure 2.6 Conjunction, thing, object and image. Author's drawing.

For example, a chair is a conjunction or a thing. It weaves together a series of qualities such as solidity, temperature, texture and colour, that we could call its sensory 'image'. The chair is also a conjunction of specific quantities – four legs, one seat, two armrests, all of a certain length, width and height – that define it as an object. When all these qualities and quantities persist in their conjunction, they create the thing we call a chair. As I said above, though, qualities and quantities cannot truly be separated, except in our imagination or vocabulary.

Every conjunction of folds weaves some folds together and leaves others out. This is what gives each thing its relative difference from other things. For example, a chair is a relative difference between all the qualities and quantities it repeatedly conjoins and all the ones it does not.

A wooden chair might have a conjoined quality of solidity but not the quality of 500 degrees in temperature or nine legs. The list of qualities and quantities that the chair does not iterate is much more extensive than those it conjoins.

Once a thing has one or fewer qualities, it ceases to be a thing. This is because for a quality to appear as a quality *of a thing*, it must appear as a quality *of* at least one other quality. If not, then it is not actually a composite or conjunction. If a quality is just a quality and not a quality *of something*, it is not a thing. A quality without a thing is like an ephemeral noise or streak of colour from nowhere.

The same is true of objects that have not conjoined with other objects. Just as a single quality is like a pop of sound or flash of colour without persistence or apparent source, a single quantity is like a like partial object or faint fog whose unity is uncertain and that disappears immediately.

The key idea in all this is that things are not discrete substances but are metastable processes supported by ongoing movements that run through them like a stream. For example, living organisms are only relatively stable pools or folds in a continual flow and energy transformation. Energy flows from the sun, conjoins in organisms, and dissipates when it dies. The folds of life are only woven eddies in the kinetic stream of energy.

Even the inorganic bodies of minerals are nothing more than relatively stable combinations of folds in the continual transformation of kinetic energy. Igneous, sedimentary and metamorphic rocks are simply three relative stages of a constant mutation and conjunction of the earth's liquid body – the rock cycle. Solid, liquid, gas, or ice, water and air are the three relative stages of a conjunction in the earth's fluid body – the water cycle.

At the microscopic level, all organic and inorganic objects are conjunctions of smaller objects with even smaller objects, and so on, all of which are in constant motion. Flows of molecules, particles and subatomic particles are continually moving and conjoining with one another.[24] Quantum waves ebb, flow, conjoin, disjoin and fold into particles on the luminous shores of existence. Even at the macroscopic level, all these objects do not produce a final state. Everything moves through our accelerating universe at incredible speeds. The point is that since all things are products of kinetic conjunction, they are technically metastable and not discrete. Flows support things by running through them.[25]

The so-called 'essential' qualities and quantities such as extension, volume and shape that define things such as chairs and organisms are contingent products of metastable conjunction.[26] When we ask philosophical questions such as 'What is the essence of the chair, or life?' we are looking for a static and unchanging answer that does not exist. Things do not have essences, just conjunctions. We look around at the world and see forms, and mistakenly think that such figures must have pre-existed some singular object like the chair I am sitting in.[27]

In other words, without constant kinetic conjunctions of folds woven together, there would be no things. There would be only a world of fragmented jigsaw pieces without coherence or structure. There would be seemingly random changes in a degree of heat, a flash of colour, a pop of sound, but nothing would make any sense unless it cycled through these ordered conjunctions.

The flow of matter keeps things moving, and the numerical folds keep them stable, but conjunctions hold the folds together. Consider a glass of room-temperature water on your table. There is no 'thing' called 'ice' in this water. However, when you put the glass in the freezer, the cycles of its molecular oscillations begin to slow down and fold closer together. The water turns to ice when all its conjoined folds slow down together and weave a crystal lattice. Once the unique oscillations of hydrogen and oxygen slow down together below the quantitative magnitude we call '32 degrees Fahrenheit', they form the composite thing 'ice'.

Things emerge through kinetic processes, but the processes are not separate or independent from the things. Flows are the processes by which things come into and go out of existence. They are the warps, woofs and vectors by which qualities and quantities are woven, folded and unfolded.

This conjunctive process is additive, 'one by one', not something attributed once and for all like an essence. There is no single substance of conjunction. The conjunctions that compose things are, like the flows themselves, in constant motion and can always undergo a change or recomposition. Therefore, the determination of the qualities and quantities of a thing is never final because the flows that compose it are steadily escaping entropically. As a process of flows, a kinetic thing is not reducible to any fixed set of qualities or quantities conjoined at a given moment.

The first-century CE Greek biographer Plutarch once asked whether a ship that the Greeks continually rebuilt from all new materials would be the same ship or a different one.

> The ship wherein Theseus and the youth of Athens returned from Crete had thirty oars, and was preserved by the Athenians down even to the time of Demetrius Phalereus, for they took away the old planks as they decayed, putting in new and stronger timber in their places, in so much that this ship became a standing example among the philosophers, for the logical question of things that grow; one side holding that the ship remained the same, and the other contending that it was not the same.[28]

From the kinetic perspective I have offered in this chapter, we can now see a third way to answer Plutarch's question. All objects, in my view, are like ships of Theseus. Everything continually flows, folds and unfolds because all objects are material and subject to ongoing decay and conjunction. We cannot say that the ship of Theseus is the 'same' ship or a 'different' ship because there never was a static form of the vessel.

Like all objects, the ship is an assembly of material flows folded up into conjoined cycles that periodically sustain its metastable form. There is no such thing as identity and therefore no such thing as difference, if 'difference' is a self-identical thing distinct from another self-identical thing.

Conclusion

This chapter has tried to show how objects emerge when flows of matter fold or cycle into metastable patterns. These folding patterns create qualities where they intersect with themselves, and quantities or 'number' where they trace a spatial and temporal loop. Instead of thinking about number and quantity as an arbitrary mental construction unique to humans or eternal Platonic form, I have tried to offer a compelling materialist alternative based on movement. Finally, I tried to describe how these cycles of folding could weave together to create the metastable composites we refer to as sensible things in the world.

Hopefully, this concept of the fold tells a plausible story of how objects are fundamentally in motion but can also appear relatively stable. However, our kinetic theory so far still does not explain how these composite objects could sustain their *order*. We are also still lacking a theory of what we call scientific knowledge, the order or composite pattern of objects. This will be the task of the next and final chapter in our kinemetric theory of objects.

Notes

1. Even imaginary numbers are hypothetical abstract quantities modelled on real quantities and are therefore derived from the basic theory of number.
2. We could look to theories of quantum gravity for support of this point, but more experimentally demonstrated is the fact that the big bang, cosmic expansion and acceleration all assume that space and time are *in motion* and emerge in and through movement.
3. George Berkeley, 'Letter to Molyneux' [1709], in A. Luce and T. Jessop (eds), *The Works of George Berkeley, Bishop of Cloyne, Volume 8* (London: Nelson, 1956), 25. See also George Berkeley, *Philosophical Commentaries*, ed. G. Thomas (London: Garland, 1989), entry 763.

4. Edward Vermilye Huntington and Georg Cantor, *The Continuum and Other Types of Serial Order: With an Introduction to Cantor's Transfinite Numbers* (New York: Dover, 1955), 86.

5. Michael Tennesen, 'More Animals Seem to Have Some Ability to Count: Counting May be Innate in Many Species', *Scientific American*, 1 September 2009, available at <https://www.scientificamerican.com/article/how-animals-have-the-ability-to-count/> (last accessed 4 March 2021).

6. Tobias Dantzig, *Number: The Language of Science* (New York: Plume, 2007), 3.

7. Many studies have already been done and new studies continue to discover animals and insects capable of counting.

8. Stanislas Dehaene, *The Number Sense: How the Mind Creates Mathematics* (Oxford: Oxford University Press, 2011).

9. Georg W. F. Hegel, *Hegel's Philosophy of Nature: Being Part Two of the Encyclopedia of the Philosophical Sciences (1830), Translated from Nicolin and Pöggeler's Edition (1959) by Arnold V. Miller, and from the Zusätze in Michelet's Text (1847)* (Oxford: Oxford University Press, 2007), 39.

10. In this sense, we could say that the smallest number is already three following Niomachus's kinetic and fluid theory of number. See *Theologoumena Arithmeticae*, 8 (Ast), in *Nicomachus of Gerasa: Introduction to Arithmetic*, ed. Louis C. Karpinski and Frank E. Robbins, trans. Martin L. D'Ooge (New York: Macmillin, 1926), 117: 'Each thing in the world is one in accordance with the natural and systematizing monad in it, and again, everything is separable so far as it partakes of the dyad, connected with necessity and matter; wherefore first their congress produced the first multitude, the element of things which would be triangle whether of magnitudes or numbers, bodily or bodiless. For as rennet curdles flowing milk by its peculiar creative and active faculty, so the unifying force of the monad advancing upon the dyad, source of easy movement and breaking down, infixed a bound and a form, that is, number, upon the triad; for this is the beginning of actual number . . .'

11. See Steven Strogatz, *Sync: The Emerging Science of Spontaneous Order* (London: Penguin, 2004), ch. 2.

12. See Hegel, *Philosophy of Nature*, 28–40. See also Thomas Nail, *Marx in Motion: A New Materialist Marxism* (Oxford: Oxford University Press, 2020).

13. Michel Serres develops a similar theory of vortices: 'The vortex conjoins the atoms, in the same way as the spiral links the points; the turning movement brings together atoms and points alike.' Michel Serres, *The Birth of Physics* (Manchester: Clinamen Press, 2000), 16. Deleuze and Guattari further develop this under the name of 'minor science' in Gilles Deleuze and Félix Guattari, *A Thousand Plateaus: Capitalism and Schizophrenia*, trans. Brian Massumi (London: Continuum, 2008), 361–2.

14. Robert Shaw, *The Dripping Faucet as a Model Chaotic System* (Santa Cruz: Aerial Press, 1985).

15. Strogatz, *Sync*, 70–100.

16. H. Bernien, S. Schwartz, A. Keesling et al., 'Probing Many-body Dynamics on a 51-atom Quantum Simulator', *Nature*, 551 (2017), 579–84. For a popular

account, see Marcus Woo, 'Quantum Machine Appears to Defy Universe's Push for Disorder', *Quanta*, 20 March 2019, available at <https://www.quantamagazine.org/quantum-scarring-appears-to-defy-universes-push-for-disorder-20190320/> (last accessed 4 March 2021).

17. The river rolls itself up like the periodicity of an electron shell.

18. Daniel Graham, *The Texts of Early Greek Philosophy: The Complete Fragments and Selected Testimonies of the Major Presocratics* (Cambridge: Cambridge University Press, 2010), 159.62 [F39], Ποταμοῖς τοῖς αὐτοῖς ἐμβαίνομέν τε καὶ οὐκ ἐμβαίνομεν, εἶμέν τε καὶ οὐκ εἶμεν.

19. Nicomachus, *Theologoumena Arithmeticae*, 8 (Ast), trans. D'Ooge, *Nicomachus of Gerasa*, 117. See also Sarah Pessin, 'Hebdomads: Boethius Meets the Neopythagoreans', *Journal of the History of Philosophy*, 37.1 (1999), 29–48 (38–9).

20. The brain science of number sense is detailed at length in Dehaene, *The Number Sense*.

21. Gilles Deleuze, *Bergsonism*, trans. Barbara Habberjam and Hugh Tomlinson (New York: Zone Books, 2011), 74.

22. See Richard Liboff, *Kinetic Theory: Classical, Quantum, and Relativistic Descriptions* (New York: Springer, 2003).

23. For a full theory of the image, see Thomas Nail, *Theory of the Image* (Oxford: Oxford University Press, 2019).

24. For a philosophical theory of diffraction, see Karen Barad, *Meeting the Universe Halfway: Quantum Physics and the Entanglement of Matter and Meaning* (Durham, NC: Duke University Press, 2007).

25. See Anatoliĭ Burshteĭn, *Introduction to Thermodynamics and Kinetic Theory of Matter* (New York: Wiley, 1996).

26. As Hume had argued. See Thomas Nail, *Being and Motion* (Oxford: Oxford University Press, 2018).

27. See Nail, *Being and Motion*.

28. Plutarch, *The Lives of the Noble Grecians and Romans*, trans. John Dryden and Arthur H. Clough (New York: Modern Library, 2000), 14.

3

The Field of Knowledge

Science establishes itself in the object.

Gilles Deleuze[1]

Objects are metastable processes that can be conjoined and ordered into different patterns. Since objects are not static, neither are their arrangements. Their figures can shift around, as birds take turns flying at the front of their V-shaped flight order. Even though the birds change places continually, they maintain a stable organisation that increases their collective flight efficiency.

I call these metastable patterns of objects 'fields' because they are distributions spread out through space and time. If the flows of matter are like threads, and their quantitative folds are like the warps and woofs, then fields are like the fabric surfaces that reveal their intricate orders and designs. The patterned designs of the fabric are nothing other than emergent orders across many woven threads. Our world of motion has no predetermined structure, but rather tendencies to conjoin, sustain and dissolve various orders. The cosmos is like a fabric that weaves itself from itself without an initial plan or end.

But why am I calling these fields of *knowledge*? We tend to use the term 'knowledge' in ways that apply only to humans, and occasionally to some animals, but I think this is too narrow a definition. Why? Well, if matter in motion weaves humans, and matter has no knowledge, then how could knowledge emerge from non-knowledge? It is like something coming from nothing. It doesn't make sense. Even if we said that knowledge was a so-called 'emergent property' of matter this still does not justify us saying that matter does not *know* at all. This would make knowledge the product of a blind and stupid process.

Instead of a sharp divide between nature and culture or non-knowledge and knowledge, I think it makes more sense to think of knowledge along a continuum. In this case, nature must have some kind of *material* knowledge or intelligence that gets ordered into human bodies. Human knowledge, in this view, is only one instance or expression of broader material patterns. More specifically, what is 'intelligent' about matter is how it orders or weaves itself through *movement* into metastable patterns, orders and designs. For instance, even minerals have a kind of material-kinetic knowledge in how they organise and sustain their atoms and molecules in a crystal lattice.

Part I of this book aims to give a non-anthropocentric definition of objects, quantity and knowledge. Then, starting from this perspective, I want to retell the history of human knowledge as one particular lineage within the broader world of moving objects. In particular, this chapter looks at the kinetic structure of scientific experimentation broadly construed.

This does not mean that there is no difference between a sea star and a particle accelerator. It just means that there is no essential or categorical difference between them. Each is a composite of differently distributed objects with its own set of capacities and relations. The general definition of a 'field of knowledge' I am offering here is only the broadest starting point of analysis, not its end. My aim here is not just to dismantle the arbitrary dualism we often make between nature and culture. In the remainder of this book, I want to use this new starting point to understand very different kinds of objects and scientific practices in history.

Fields of objects weave the world, but these fields are not 'flat' or the 'same' in any way. These fabric-like fields are undulating and metamorphic like the waves of the ocean. My purpose in defining objects and knowledge so broadly is not to obliterate the human subject or trivialise its unique knowledge structures. Rather, I do this because I do not want to assume any 'obvious' division between nature and human knowledge from the outset. I want to show how human knowledge structures emerged historically like patterned designs on a quilt or waves on the ocean.

By attributing knowledge only to humans, I believe we have unduly kept philosophy from thoroughly interrogating the continuity between self-organising material systems and human ones. The history of science has tended to characterise matter as stupid, passive and rule-governed. This doesn't seem right to me. Matter, I will argue throughout this book, is active, receptive and creative. Matter flows from hot to cold, oscillates and responds to itself, and in so doing, creates the qualities and quantities of objects. Human bodies are made of these objects and are not independent from them.

Why is it that when a human draws a fern it is art, but when matter grows into a fern, it is not? When humans keep track of a solar day with a calendar, we call this science, but when plants do it, we call it mechanical response. Why? Plants may not write out mathematical formulas on paper with pencils, but in some sense don't they perform the most daring non-linear equations with their fractal limbs?[2]

What I think we need is a grassroots philosophical framework that begins from an expansive definition of objects and knowledge as something all of nature does. Then, from there, let's see how human scientific practices emerged. If we start by defining scientific knowledge as something only humans do, then the rest of nature inevitably falls short. In this chapter, I want to describe the basic kinetic structure of how human knowledge works through metastable patterns of iteration and circulation.

The field of circulation

A field of objects is like a fabric woven from interlocking vortices. Flows of energy continually circulate through a particular order of objects even when individual objects dissolve and new ones take their place. Our bodies continuously replace cells, like the parts of Theseus' ship, but we maintain roughly the same shape and sense of self-awareness. We still feel like the 'same' person we were a year ago, even though almost all our cells are different because our bodies are fields of circulation.

A field is a flow that weaves through a series of vortical folds in a particular order and then repeats the process to maintain the order. Objects allow flows of matter to persist through cycling, but fields of circulation allow all the objects to preserve their order over time.

Without a field, objects are like abstract disordered objects. For example, when we look into the night sky and see a planet, we might assume that this planet has no relationship to the other planets, moons or the sun. If we tracked its motion, though, we might find that it follows a steady path through the sky and sometimes reverses course and moves backwards. Why does this happen, and what is the relationship between this movement and the other planets? This is a question about its field of circulation.

The German astronomer Johannes Kepler was the first to describe this field with his 'laws of planetary motion' (1609 and 1619). What initially appeared to be erratic or random movements Kepler saw were ordered patterns of circulation. Kepler began with a fragmented list of the observed magnitudes of planetary motions inherited from his teacher, the Danish astronomer Tycho Brahe. However, with his observations of Mars, Kepler discerned a consistent and repeatable elliptical motion in all the planets. He discovered the iterating field of circulation of planetary motion.

This planetary field of circulation did not always exist but emerged historically, and is still emerging in our evolving solar system. Kepler's laws also have their own field of circulation distinct from the planetary motions. His genius was that he set up the first accurate coordination system between the planetary and notational fields.

Circulatory fields are what give objects their repeatable relations with one another. The persistence of metastable relations gave planetary motions their regularity and allowed Kepler's mathematical model to track and predict it consistently. Without the iteration of these relations, nature would be random, and human scientific knowledge would be impossible. Without fields, quantities and their composites would have no stable patterns or connections.

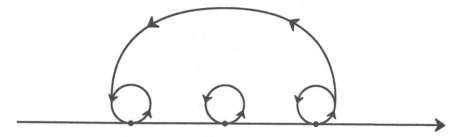

Figure 3.1 Field of circulation. Author's drawing.

Epistemology

What is knowledge? In my material kinetic theory, knowledge is the iterative coordination of objects and their relations, like an intricate piece of embroidery. In this sense, knowledge does not *represent* the world but is part of the world. It is the way the world weaves and orders itself. For example, Kepler's laws do not represent planetary motion but instead have their own movement and order that coordinates to planetary motion in a particular way. The same is true of Kepler's idea or understanding of planetary motion. His body and brain have their own field of circulation composed of neurons and electro-chemical flows that he can coordinate with the planetary movements.

In this sense, we might think of a 'kinetic epistemology' as a more performative 'know-how' that occurs through action. When we ride a bicycle, we *know how* to coordinate our body's ordered field of circulation with the bicycle's field. Conversely, the bike also knows how to move with a human body because of its objective order.

Since all objects are in motion and have some degree of action, they all know how to move in the context of other objects. In the case of know-how, the knower enters into feedback or circulation with what they know. Each time an object *does* something new, it *knows how* to do something new *at the same time*. The more object-actions can be circulated together without falling apart, the more robust and broader the range of knowledge.[3]

The difference between riding a bicycle and doing experimental science is not an ontological difference in kind but a difference in the range of circulation and coordination. Scientific experiments rely on the know-how of a whole host of objects in exact orders. As the philosopher and historian of science Andrew Pickering writes, 'Scientific knowledge should be understood as sustained by, and as part of, interactive stabilizations situated in a multiple and heterogeneous space of machines, instruments, conceptual structures, disciplined practices, social actors and their relations, and so forth.'[4]

In this sense, there is no such thing as knowledge 'of' an object because knowledge does not represent anything. Since knowledge is immanent to its objects, it can only coordinate with them. There is no such thing as knowledge that is not in an object or that floats above history. Every object is always in an 'interactive stabilization', as Pickering says. Knowledge is always in a distribution of objects in the whole 'practical setup necessary to render the fact visible', as the philosopher of science Bruno Latour describes it.[5]

As an example of this, Latour gives a beautiful description of the difficulty of how French scientists 'found tuberculosis' in the mummified body of Ramesses II. What French scientists found in Ramesses II's body, according to Latour, was not an ahistorical object called 'tuberculosis'. Instead, these scientists found a metastable field of circulation that coordinated ancient tuberculosis to modern tuberculosis. The tuberculosis object they found only appeared as an object in an ordered distribution with other objects. We cannot separate modern tuberculosis from its historical and material appearance alongside objects such as laboratories, hospitals, specialists and X-rays. Since ancient and modern knowledge fields are different, this is not the 'same' tuberculosis object, just as Theseus' ship was never the 'same' ship. However, ancient tuberculosis bears a similar enough pattern to modern tuberculosis that we can coordinate the two into a single field of knowledge and say, 'Ramesses II died of tuberculosis.'

In short, the material, technological and social conditions of scientific knowledge are inseparable from the performance of knowledge itself. All the objects of the field are part of its knowledge. Knowledge, therefore, is not one object among others. It is the immanent kinetic coordination of ordered objects. This is not social constructivism because there is no separate object called society that is constructing knowledge.

As a piece of fabric is nothing but its woven threads, fields are neither just objects nor just the relations between objects. The field is the whole persistent immanent pattern. The field of circulation is the continual flow that traverses things, binds them together and orders them. It is the way things always move together relative to one another. The path of circulation is not a pre-existing order waiting for someone to fill it with objects. Objects and their order co-evolve and co-emerge because the whole process – flow, fold, field – is in indeterminate but metastable motion.

The theory of fluxions

Everything moves, not as a totality, but as an open and indeterminate process. This means that there is no single immobile point from which to measure the complete order of things. Motion and its measure, as Einstein stated, are both relative.

'The theory of fluxions' provides a logic for understanding the relative ordering of things in a field of circulation. A 'fluxion', as I define it, is a degree of motion relative to the movement of its field. Fluxions are the degrees and relations of coordination between objects. An object moves 'more' or 'less' with its neighbouring objects and relative to the motion field that orders them.

For example, relative to one object, another appears to be moving, but relative to another, the same object appears still. The French philosopher René Descartes described the motion of a ship in his *Principles of Philosophy*. Relative to the shoreline, a passenger on a vessel appears to be *moving* down the river. Relative to the ship, the same passenger is *not moving* relative to the neighbouring bodies aboard the ship. However, if the ship was being pulled by the wind upriver at the same speed that the current was pulling it downriver, it would *not be moving* relative to the shore but *would be moving* relative to the changing wind and water surrounding it.[6]

The point is this: everything is moving but only more or less, *relative* to other motions. I call a 'fluxion' the relative kinetic difference between objects that defines their order in a field. The field of knowledge is the flow or continual function, within which the fluxions relate to one another.[7]

Each object has a periodic motion but is also related to other objects in the same field. Every field of motion is, therefore, composite. For example, Descartes continues, if you are walking along the deck of a ship with a pocket watch, the watch's wheels have their motion. However, added to this motion is the motion of your body along the deck. Added to your body's motion is the motion of the ship tossed about on the waves. Added to that motion is the ocean's motion as a whole. Added to this is the motion of the whole rotating earth, and so on.[8]

All of these motions are part of the same field of motion. However, each also moves more or less relative to the others. Relative to the watch, your walking body is less mobile, and relative to your walking body, the ship is less mobile as you walk across it. Still, relative to the ship, the waves are less mobile as it sails across their surface, and so on in relatively *decreasing* degrees of mobility or fluxion. From the inverse perspective, the waves move across the ocean's surface, the ship moves across the surface of the waves, and so on in relatively *increasing* degrees of mobility or fluxion.

Einstein's theory of special relativity's significant contribution to the idea of coordination was to show that space and time themselves are relative to one another *in motion*. The field of motion is what relativises space and time. As the degrees of fluxion increase, time slows down or 'dilates' and space 'contracts' relative to a given field of coordination. Therefore, every degree of flux also determines a degree of spacetime, not the other way around. Following special relativity, the theory of fluxions offers a description of spacetime grounded in the motion of objects. A fluxion is simply a difference in the degree of motion relative to a given kinetic field. On the other hand, a kinetic field is a *relatively* immobile background from which we measure the different fluxions. Therefore, a field is that which has zero motion relative to the rest of the degrees of motion.

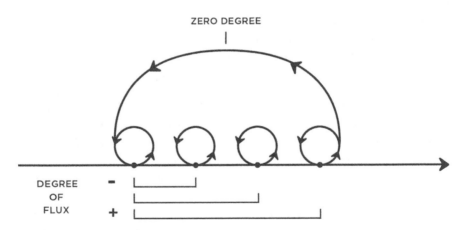

Figure 3.2 Fluxion. Author's drawing.

The protein-folding problem

Let's bring our three key concepts of 'knowledge', 'coordination' and 'fluxion' together in a concrete example from biochemistry. In 1962 Max Perutz and John Kendrew won the Nobel Prize in Chemistry for their pioneering work in determining the structure of globular proteins, such as haemoglobin in red blood cells.[9] However, while their work laid the foundations for structural biology, it also raised a more fundamental question of how protein structures are produced in the first place, known as 'the protein–folding problem'.

Progress has been made in this field,[10] but one of the reasons it remains an unsolved problem today is that the flow of amino acids that produce these proteins is fundamentally stochastic. Amino acids flow pedetically, twisting and intersecting with one another, and then eventually make a 'random coil'. This 'random coil' is then further folded over itself into various ordered three-dimensional protein structures. Depending on these folds' structures, these proteins perform essential tasks of the cell, including DNA replication. The kinetics of this 'folding pathway', like all flows, moves from higher- to lower-energy structures.[11] Biochemists have now collected more than 80,000 protein structures in atomic detail in the Protein Data Bank.

However, there is a second protein-folding problem. Sometimes proteins are 'misfolded' or only partially folded into highly disordered aggregates or 'granules'. These inactive protein structures can interact with one another and form larger insoluble aggregates that cause degenerative disorders and cell death.[12] Scientists have linked such degenerations to Alzheimer's, Huntington's and Parkinson's diseases.[13] A recent research line has discovered that adding RNA to the aggregates slows down the disordered aggregation and can even prevent it from occurring.[14]

However, before we can do a kinetic analysis of this scientific field, we need to understand how biologists made this determination empirically through the scientific experimental process. The process is quite complex, so I will stick to only the major moves, which are not greatly different from the primary activities of the experimental sciences more generally. This is how modern science performs the process of experimentation, and I want to look at each of the kinetic aspects of this practice here.

First, scientists remove a slice of brain matter destroyed by the neurodegenerative process of misfolded proteins. Second, they isolate the protein and use its genetic structure to reproduce a collection of similarly misfolded proteins outside the brain. Third, they place the duplicated proteins in a growth medium and track their aggregation rates, growth or dissipation using a light-scattering device that measures the molecular weight of the proteins. Fourth, they try adding different replicated RNA chains at different protein aggregation stages and record the differences in molecular weight with the light-scattering device. Fifth, they record and synthesise their data in visual graphs, narrativise the negative correlation between misfolded protein aggregation and messenger RNA, edit their paper, and send it out for publication. We could include numerous other intermediate steps, but the broad strokes will be enough to test our theoretical framework so far.

Kinetic analysis

We can now examine each of these steps, in turn, as coordinated degrees of fluxion in a field of scientific knowledge. The steps below are specific to the example above, but its essential components are common to almost all experimental and applied sciences.

The sample

The first object I want to look at is the slice of neurologically degenerated brain. Experimental science, including natural and social sciences, always begins with some bit of matter or relatively raw material from which it builds and to which it must eventually return to explain. This bit of raw material has its qualities and quantities unique to it, and thus its own kind of knowledge or 'know-how', in the sense I described above. This slice is different from other brain slices of the same brain and different from other degenerated brains. Since one cannot experiment on all brains, scientists select one slice of one brain at a time, with the hope, not a guarantee, that other selected brains could support the same epistemological field of coordination.

This is the function of the scientific 'sample'. From the vast field of matter, scientists select an object or objects to enter into a field of knowledge. They are not representations of nature but a baseline or working presupposition of repeatable coordination relative *to a field of knowledge*. If each brain is unique, then accurate representation is technically impossible. Two slices of brain are never identical, especially given the dynamic complexity of the misfolding process.

Knowledge is always regional, situated, historical and bound to its material field of coordination. There is no scientific knowledge independent from its distribution in the concrete field within which it is ordered. Therefore, the sample can never represent the world; it can only function as the first object in a series. Insofar as it moves, it has its own know-how and functions as the smallest non-zero degree of knowledge in the field. The sample is the baseline from which all subsequent degrees of motion will depend. The whole network of hospitals, surgeons, laboratories, machines, sterilisation equipment and so on that go into collecting and shipping these samples conjoin with this sample object. Even if scientists do not credit them in the published paper, they are still material kinetic aspects of the sample.

Extraction

The second object in this field is the 'isolated', 'extracted' or 'purified' object. The sample itself is already a kind of extraction because it is a selection taken or chosen from a vast material flow network. From any sample of matter, there is always more than one thing going on. From the raw sample, one or more other objects are extracted or focused on, often at the expense of excluding other aspects of the sample.

In our example, the misfolded protein aggregates are extracted from the brain tissue and synthesised in a laboratory. The proteins synthesised in the laboratory are not the same as those in the sample. The sample, too, was not the same as other slices in the brain. Therefore, the newly synthesised lab proteins do not represent the sample proteins but coordinate with them in the same field. More specifically, their relationship is one of *kinetic dependence* or fluxion. Without the former, scientists could not have synthesised the latter. The kinetic action of the latter depends on an extraction and relative immobilisation of the former.

Included in this second object, of course, is a different but no less complex set-up. A laboratory, trained workers or students who know how to use the machines, expensive machinery to shake giant tubs of *E. coli* bacteria in which the proteins grow, and a ton of 'failed' or unstable versions of the proteins are all part of the field. These motions rely on and presuppose the initial sample, but what they produce and how they produce it does *not represent* the sample. It is an entirely different complex kinetic object with its own conjunction of sub-objects. The extracted object is thus *coordinated* in a relation of kinetic dependence with the sample by one degree of flux or kinetic separation from the relatively much less mobile and worked-upon sample.

Measurement

As we have said, all matter is already self-measuring in the sense that it has a duration and magnitude. Knowledge, however, is the ordering of measures and magnitudes with one another. For example, a protein has a magnitude expressed in the duration and extension of its folded body. When a laser shines on this body, the light diffracts or scatters at various angles depending on the aggregate's molecular density and shape. This is called Mie scattering. Scientists can use the resulting diffraction to determine the object's molecular mass, based on the known diffraction patterns of specific molecules.

All measurement is thus itself a kinetic act. First, it requires a degree of kinetic dependence or fluxion on the object measured. Without this dependence, the act of measurement has no meaning. Measurement depends on what one is measuring. Therefore, measurement is always relational between objects and produces a result that describes their coordination. The measurement tool itself must, therefore, be another object included in the 'measurement'. Without a measuring device, there is no measurement. Thus, every measurement presupposes a third object coordinated with the measurer and the measured. We must also consider the lab assistant, the lab, the tools needed for scattering and measuring light, the extracted object and the resulting numbers as objects in the measurement object.

In other words, neither the lab worker, the laser scatter nor the numbers bear any resemblance to the protein aggregate – they do not represent it. Instead, they coordinate with it in a specific relationship of kinetic dependence. Importantly, this is not just a single coordination but also a repeatable coordination of dependency between relatively standardised objects and procedures that scientists can consistently perform without a significant variation in the relation.

A measurement is like a recipe for a field of knowledge. Science works very hard not to produce just any kind of coordinated dependence but a stable set of coordinations and coordinated objects that will allow the field of knowledge to persist and expand. This is in part because a field of knowledge does not end when a paper is published – it circulates. For the knowledge to endure, it must be reproducible across time. Once the field's flow reaches its limit, it must be able to return to its sample and other samples of nature and produce a similar distribution of fluxions in similar relations of dependence. The creation of standardised units of measure provides instructions about the specific kinetic relations one created in the field.

So, measurements do not technically represent anything. Instead, they are simply a shorthand for a detailed description or set of instructions to create a particular composite object (measurer, measured and measuring device). In other words, a measurement is a way of saying, 'If one uses this tool in this way on this object under these conditions, the coordination should produce this number on the device.' If not, the coupled oscillations of the field's folds will enter into a different coordination of fluxions.

Variability

The fourth object in this field is the variable object. By introducing a variable object, such as RNA, into the growth medium and re-measuring all the misfolded protein granules, one can now record a change or difference in their measurement.

A variable object is a differential object composed of two different sets of measures. One measure is of the protein before and after, and the other is of the RNA before and after. In this sense, the variable's status depends on the previous measurement of the protein without the RNA and a resulting contrast. In a linear and retroactive sense, the measurement, extraction and sample all now appear to be 'dependent' on the newly added 'independent' variable. Once the scientists introduce RNA, misfolded proteins begin to slow their growth. But the apparent linearity and causality of the variable, presupposed in the distinction between dependent and independent variability, comes only retroactively at the end of the experimental sequence.

However, the RNA kinetically depends on all the previous objects in the field. All the prior objects of the field are the condition for the so-called independence of the variable. Kinetically speaking, then, we must say that the RNA cannot be the cause of a change in the protein extraction without the protein extraction causing or allowing itself to be slowed down by the RNA.

The interaction is thus collective and kinetically performative. We are justified in saying this because there is a mutual kinetic action in the protein and the RNA as they interact. They are dimensions of the same variable relation. We will say more about complex kinetic causality in a moment. For now, the critical point is to see that the variable object is far from being a representation of the extraction but rather depends on the whole field. The variable requires not only a sample but an extraction of the sample and a measurement of the extraction as the conditions for its own differential measurements.

Recording apparatus

The fifth and final object in this field is the recording object. Modern science tends to privilege the narrative and conceptual object of the epistemological field: the published paper. This is because it is usually the final object in the series that includes a relation to all the previous coordinated objects on which it depends. In this recording object, we find a host of other objects, including group discussion among collaborators, teacher–student mentoring, computer software programs for graphing and producing text, and a vast sociological network of peer reviewers, journal editors and so on. In short, we find all the material and historical objects that go into organising data, narrativising knowledge and publishing the results.

But a paper is still not a representation of anything. It bears no resemblance to any of the laboratory objects or synthesised proteins. Narrative writing and graphs have a long history of producing established coordinations between spoken words, letters and numbers, which we will explore more in Part II. Because of this, we accept written words and images as coordinated with other objects by historical convention. The scientific article or book is, therefore, a shorthand recipe or instruction manual for reproducing and expanding a coordinated distribution of kinetic objects. It is not a universal truth.

The scientific paper says that if you want to do what we did, here is how you distribute and order your field. Here is the order of operations of the field, its possible connections to other fields, and how we might enlarge it, preserve it, apply it medically and technologically, beyond its currently limited region. In the theoretical sciences such as theoretical physics or neuroscience, however, the recording apparatus is almost the entire process. There is no material sample, and no extraction or variable to test (although perhaps other experimental studies could be referenced). Number and narrative are the means and end of the theory because a proof is a well-ordered recording apparatus. A conclusion is ordered by the well-founded conditions of its own starting point. It is a solution to its own problem or a recipe without ingredients. In theoretical papers, there are also no representations. There are only rules or instructions for the coordination of inscriptions.

The experimental recording apparatus such as the peer-reviewed journal article, of course, also has its own numerical and linguistic distribution of numbers, letters and graphs, which we coordinate internally. Moreover, they are dependent on objects outside the article. Instead of representing the field, the article is simply one more object in the field that depends kinetically on a certain coordination of all the other objects in the field. The article allows one to return to the first object in the field to reproduce and expand it.

What is observation?

Instead of asking what kind of thing something is, perhaps we should ask about its objects and their order. What are its degrees of kinetic dependence or fluxion, its limits, and its range of expansion? We can even ask about the qualities that coincide with the quantities, as in the qualitative sciences. The field of circulation, therefore, entails a very different theory of observation.

If knowledge is kinetic and performative, then observation is not a representation of a state of affairs. Instead, observation is a kinetic act that coordinates the objects it observes in the field. As such, observation both affects and is affected by the field of knowledge. Observation is something that a field of knowledge *does*. It is a process that preserves the regional stability of its joint degrees of fluxion.[15] Observation and preservation both come from the Latin word *servō* meaning to 'watch, maintain, or keep safe'. Watching is related to maintaining and is not a neutral view on things.

We tend to think of humans as the only ones who can observe, but this is not the case from a materialist perspective. For instance, in quantum physics, objects can 'observe' other objects by affecting them with photons or other particles.[16] Observation is, therefore, not only a mental act but a material act of self-sensation and self-preservation.

Every object in a field of knowledge observes and preserves other objects in the same field by reproducing, again and again, the specific order of dependence or fluxions. Each object not only observes itself by affecting itself. It also observes other objects by affecting them as well. Observation is a process by which a whole field of objects reproduces a relatively stable order.

Observation is like the kinetic synchronisation of objects that occurs in the beautiful twisted scroll waves of the Belousov–Zhabotinsky chemical reaction, discovered by Arthur Winfree.[17] This is a non-linear pattern of chemical oscillation where changes in some molecular objects trigger changes in others repeatedly in a series without equilibrium.

If there were no collective observation or preservation, each object in the field would be unrelated, unordered and isolated from the others. Objects would not sync in any order with their neighbours, and the field would dissipate. Observation is the preservation of self-relations among objects in a field.

This does not mean, however, that observation always reproduces the same field. This is why fields mutate, change, expand, contract and so on, precisely as in the Belousov–Zhabotinsky reaction. Since observation is a kinetic process, it is always changing. However, it is always changing in metastable patterns of coupling and synchronisation. From fireflies to heartbeats, to ecological populations, to the experiments that study them, fields are continually mutating in more or less non-linear metastable formations.

The problem of scientific realism is thus poorly posed as a strictly human epistemological problem: How do humans know the true nature of objects in themselves? If 'in itself' means an isolated object without relation, there is no such thing. However, if 'in itself' means not dependent on a human field of knowledge, then, of course, there are objects and fields of knowledge in themselves. Matter is self-observing; it does not need a human to observe it to be known or exist.

Observation preserves and supports the structure of relations among coordinated objects without a central authority or human observer. When humans or other objects are added or taken away from a field, the field changes. Objects change when humans observe them not because of some spooky epistemological action at a distance, but because of the real kinetic action of light and motion itself. Shooting photons of light at something to see it changes it. Fields change when something new joins it, whether this is a microscope or an airborne fungus. In this way, the field becomes different than it was because of the introduction of new observers.

Conclusion

This chapter concludes Part I of this book. In it, I have introduced three concepts to help us make sense of how kinetic objects emerge, change, combine and dissipate. Objects, in my view, are not static, and neither is their order. Instead, I have tried to provide a process-philosophy of objects and their relations as material and metastable processes. This non-anthropocentric approach involved expanding our typical definitions of number, knowledge and observation, with significant consequences for how we think about human scientific practice. The history of science is one expression of material history and not a representation of it.

However, the concepts of flow, fold and field are highly synthetic since I have had to extract them from their historical context here. I hope that they will give the reader a more 'kinetic' vocabulary to think about objects in this book. This is important because although matter has always been in motion, the history of science has worked very hard to obscure or explain the primacy of motion by something else. Scientists accomplished this with all kinds of ingenious methods. In particular, in Part II, I would like to look more carefully at four key historical fields of circulation and the types of objects they have created. My unique contribution to this history, however, is to show the hidden fields of circulation and movement that pervade scientific practice.

Notes

1. Gilles Deleuze, *Letters and Other Texts*, ed. David Lapoujade, trans. Ames Hodges (New York: Semiotext(e), 2020), 289.
2. See Colin Tudge, *The Secret Life of Trees: How They Live and Why They Matter* (London: The Folio Society, 2008).
3. See Davis Baird, *Thing Knowledge* (Berkeley: University of California Press, 2004), for an interesting theory and history of materialist epistemology. The author describes three types of thing-knowledge: Model Knowledge (representation, explanation, accurate), Working Knowledge (performs regularly and reliably) and Encapsulated Knowledge (model plus working). Different objects have different knowledges.
4. Andrew Pickering, *The Mangle of Practice: Time, Agency and Science* (Chicago: University of Chicago Press, 1995), 70.
5. Bruno Latour, 'On the Partial Existence of Existing and Non-existing Objects', in Lorraine Daston (ed.), *Biographies of Scientific Objects* (Chicago: University of Chicago Press, 2005), 250.
6. René Descartes, *The Philosophical Writings of Descartes: Vol. 1*, trans. and ed. John Cottingham, Robert Stoothoff and Dugald Murdoch (Cambridge: Cambridge University Press, 1985), 228; Part II, Section 13.

7. Isaac Newton, *A Treatise of the Method of Fluxions and Infinite Series: With Its Application to the Geometry of Curve Lines* (London: Printed for T. Woodman at Camden's Head in New Round Court in the Strand, 1737).
8. Descartes, *The Philosophical Writings of Descartes: Vol. 1*, 236; Part II, Section 31.
9. J. C. Kendrew, G. Bodo, H. M. Dintzis, R. G. Parrish, H. Wyckoff and D. C. Phillips, 'A Three-dimensional Model of the Myoglobin Molecule Obtained by X-ray Analysis', *Nature*, 181.4610 (1958), 662–6.
10. Ken A. Dill and Justin L. MacCallum, 'The Protein-Folding Problem, 50 Years On', *Science*, 338.6110 (2012), 1042–6.
11. P. S. Kim and R. L. Baldwin, 'Intermediates in the Folding Reactions of Small Proteins', *Annual Review of Biochemistry*, 59 (1990), 631–60.
12. T. K. Chaudhuri and S. Paul, 'Protein-misfolding Diseases and Chaperone-based Therapeutic Approaches', *The FEBS Journal*, 273.7 (2006), 1331–49.
13. F. Chiti and C. M. Dobson, 'Protein Misfolding, Functional Amyloid, and Human Disease', *Annual Review of Biochemistry*, 75 (2006), 333–66.
14. Erich G. Chapman, Stephanie L. Moon, Jeffrey Wilusz and Jeffrey S. Kieft, 'RNA Structures that Resist Degradation by Xrn1 Produce a Pathogenic Dengue Virus RNA', *eLife*, 3 (2014), published online 1 April 2014, available at <https://elifesciences.org/articles/01892> (last accessed 4 March 2021).
15. The word 'observe' comes from the PIE root *ser*, 'to serve, keep and protect'.
16. 'Knowing is a matter of part of the world making itself intelligible to another part.' Karen Barad, *Meeting the Universe Halfway: Quantum Physics and the Entanglement of Matter and Meaning* (Durham, NC: Duke University Press, 2007), 185.
17. Steven Strogatz, *Sync: The Emerging Science of Spontaneous Order* (London: Penguin, 2004).

PART II

A HISTORY OF OBJECTS

I THE ORDINAL OBJECT

4

The Centripetal Object

If objects are created and sustained through movement, then these patterns also have a history. As we saw in Part I, objects can distribute themselves into different metastable orders I called 'fields'. History is full of these fields. Part II of this book tells the history of how four major fields of objects emerged from prehistory to European modernity.[1] More specifically, in Part II I want to show the history of the emergence of these four kinds of objects from a movement-oriented perspective.

I do not want to merely catalogue the invention of objects or summarise what people said about them. I want to illuminate the material agencies and patterns of mobility that produced and sustained various fields of objects. Instead of thinking of objects as static forms or human constructs, I want to think about them as material and kinetic processes. In particular, I want to look at the history of how the sciences have created and sustained them.

If the world is full of objects, why look only at the history of science? Well, what is science? I define 'science' quite broadly as the creation and ordering of objects *as quantities*. In principle, objects are inseparable from their qualities, relations and modes of being. There is no such thing as a quantity without a quality or a relationship to something else. However, in practice, particular groups of humans have developed a long tradition in which they act *as if* particular objects were *mostly* or *only* quantities.

These abstractions have a deep and fascinating history, including a wide range of more and less abstract objects. Treating things as quantities to various degrees makes all kinds of new fields of knowledge and technology possible. However, it also obscures some other elements. For better or worse, science is the human practice of focusing on the quantitative dimension of things to make and arrange objects in new ways. The arts tend to do something similar by concentrating intensely on the qualities of things, and politics by focusing on the relations between things.

Science then, in my admittedly broad view, includes everything from the creation of Palaeolithic tally marks to contemporary biochemistry. It covers technology, mathematics, logic, as well as the natural and social sciences. By retelling the long history of science from the perspective of motion, my aim is not to describe everything it leaves out, but to offer a more accurate description of how it works. The history and philosophy of science are too often seduced by the operative fiction that science works with abstract quantities. Instead, I want to show how such seemingly discrete quantities emerged from iterative movements and patterns. Among other things, I also want to show what role the agency of objects plays in this process.

I have organised Part II into four sections. I dedicate each section to one time period in history and the main kind of object that emerged and dominated during that time. At the beginning of each of these historical sections, I have written a short introductory chapter that defines the primary type of object under consideration and the pattern of motion that produces this type. I wrote these four introductory chapters as more synthetic or conceptual works after I did the historical research for the later chapters. My purpose with these short chapters was to give the reader the basic idea or broad strokes of the historical chapters in advance. That way, they would see how and why I presented all the historical details as I did.

I could not write about every science in history, so I tried to pick at least one notable instance from logic, mathematics, technology and natural science from each period. The goal of each of these historical sections was not to be exhaustive but to be synthetic. I want to show that many different scientific practices during a given time all created and maintained a unique type of object by following a similar pattern or field of motion.

After the short introductory chapter at the beginning of each section, I wrote two richer and more historically detailed chapters, except in the first section, where there is only one. These chapters are not necessarily organised chronologically but are organised topically around several significant scientific practices of the time, such as the invention of 'numerals', 'dynamics' or 'infinity'. Since these chapters deal with the historical details of science, they are, by nature, more technical in style. I have tried my best to explain all the terms without assuming any specialised knowledge from the reader. However, given the breadth I have wanted to cover, this was a challenge.

Finally, before launching into our first introductory chapter, I would like to remind the reader why this history is so critical. There is not only one kind of object. There are several kinds of objects that all depend on the processes that produce and reproduce them. It has taken thousands of years for all four of these kinds of objects and their fields to emerge, and they have not disappeared. The objects we hold in our hands today are mixtures or hybrids of these four historical types. To understand any given object of contemporary life, we have to understand something about the general patterns that first produced them and that still produce them. Today's objects are different, but the patterns of motion that make and maintain them are not. Identifying these common historical patterns allows us to make theoretical use of them today to understand contemporary objects. This is what I will try to show in more detail in Part III.

But now, let's move on to discuss the first kind of object to emerge in human history and the kinetic pattern that sustained it.

The ordinal object

The first kind of scientific object emerged during the Palaeolithic period (3.3 million–10,000 years ago) and achieved its zenith during the Neolithic period (10,000–5000 BCE). I call this earliest human object an 'ordinal' object because it is part of a sequence or series of objects.

When we count objects, we tend to use numerals such as 1, 2, 3 and so on, but this is not the only way to count. For the vast majority of human history, there were no numerals at all. Early humans made, ordered and counted objects sequentially, 'and then, and then, and then . . .' or 'first, second, third . . .' The German mathematician Georg Cantor introduced the term 'ordinal number' into mathematics in 1883, but the broader usage of this kind of object is much older. The use of ordinal objects goes back to the birth of prehistoric science and is at the foundation of all other orders of objects.

What is particularly fascinating to me about the invention of ordinal objects is how early humans coordinated them together with one another. By doing so, they not only put things in spatial and temporal order but connected the sequences to generate composite series.

How did humans start organising the world into woven ordinal series? Some scholars attribute this innovation to the evolution of the human brain. This seems relevant, of course, but only pushes the question back one step further. I still want to know what the initial material changes were that led to this specific brain evolution towards ordinal counting. Furthermore, I think we have to admit that the manipulation and coordination of ordinal objects certainly requires more than a change in brains. So, I think the evolutionary hypothesis does not quite explain what I want to know.[2]

Most other scholars, however, have committed a worse error in thinking by far. When asked how humans first began counting, they say that early humans probably starting counting on their fingers, 1, 2, 3.[3] Although this sounds intuitive, it is historically impossible. The cardinal numbers,

1, 2, 3 etc., were not invented until 3500 BCE.[4] This is an excellent example of the fascinating hybridity of objects. Once they emerge, they tend to feel as if they are entirely natural and have always existed when they haven't.

Ordinal objects, though, are not the result of evolutionary biology or even human embodiment alone. They are the result of historical and material processes that are the subject of this chapter. In particular, this chapter and the next argue that ordinal objects emerged and propagated in prehistory in three main ways: centripetal motion, sequence and one-to-one coordination. Here, I offer only the very broadest strokes of these processes and how they work. In the next chapter, though, I describe how these processes spread in various prehistoric scientific practices. Before getting into the history, let's now look at each of these three processes in turn.

Centripetal motion

The first kinetic process required for the creation of ordinal objects is centripetal motion. What is centripetal motion? In my usage here, centripetal motion is any movement from the periphery towards a central area. For any field of objects to be set in some order, they must first gather together somehow. Centripetal motions are what collect a diverse range of objects from various places towards a central area.

We can see this movement in many aspects of nature. For example, honey bees gather pollen from a far-flung periphery of flowers and bring it back to their hive. Once they have centripetally collected pollen at the hive, the bees can begin to order it into hexagonal honeycomb cells. Birds and other nesting animals gather sticks, moss and grass from their periphery and bring it back to their central nests. Even the human body moves quite naturally in this way. To survive, it must move about some extended periphery to gather food and water and then direct it into its central cavity.

By this simple pattern of motion, a sequential order begins to develop. Some things are pulled to the centre and piled there before others. One after another, a bird piles first larger twigs, then smaller twigs, then some dried leaves, in order. The very act of gathering several things necessarily entails a kind of a sequential order as each is deposited in the centre. Centripetal motion is, therefore, a necessary kinetic precondition for any ordered field of objects.

The order of ordinal objects depends intimately on their position alongside other objects and a relatively stable centre that supports them. However, if the centre keeps changing faster than the objects ordered on it, the order will fall apart. In this way, the order of objects assumes the mobility of objects and a relatively stable background field. The differences between objects are not pre-given but build up through centripetal movement.

Therefore, the existence of ordinal objects also depends on a relative kinetic difference between a stable background and a mobile foreground of variously ordered objects. For example, a bird gathers twigs into a nest and then deposits her eggs in the nest. The amount of eggs is 'more or less' relative to how many twigs there are and the size of the nest.

The kinetic pattern of inscribing a tally mark on a piece of bone is also centripetal. The first kinetic event that made this possible was the transition to bipedalism in human animals. By walking upright on only two legs, humans had much freer use of their arms, hands and mouths. Humans could gather more and larger objects together and more easily arrange them with their freed-up limbs. An animal bone and a sharp rock, for instance, are two very different shaped and sized objects that are much easier to carry in one's hands than in one's mouth. These are the two objects needed to make a lasting mark or notch.

After humans gathered the bone and stone in the same place with their hands, they combined them by a similarly centripetal motion. The bone acted as a relatively stable field for ordering and relating the marks left by the stone. It is the relatively smooth and unmarked bone surface that allowed the notches to be preserved and distinguished from one another in order. The rock pressed down on the bone, dragged along the surface, picked up, then repeated. Each stroke moved in a centripetal cycle out and back again.

In these ways, and others discussed in the next chapter, ordinal objects emerged through centripetal motion.

Sequence

The second method for creating ordinal objects is the use of sequence. As we incrementally gather various objects from a periphery towards a centre of accumulation, the process establishes a series of spatial and temporal differences between objects. We inevitably collect some objects before others. Other still are not gathered at all and remain on the periphery.

For example, the simple centripetal act of gathering firewood into a pile creates a sequential order of objects. The first sticks that one gathers will be at the bottom of the pile and towards the centre. The later sticks will be piled on the top and roll off towards the outside of the pile. The pile grows larger as sticks accumulate and smaller as they dissipate. Objects are always ordered *sequentially* in space as 'less or more', and in time as 'before or after'.

However, it is also crucial that the first object in the centripetally gathered series persists. Only when the first object endures can there be a difference between it and subsequent objects. In other words, the first object only becomes a 'first' for the second that follows it.

This is the paradox of the first object in a sequence. The first object cannot be the 'first' in a 'sequence' if it is the *only* object. The 'first' object is only retroactively first, once there are subsequent objects in the sequence. However, the 'first' centripetal object is also crucial because it defines the difference between gathered and ungathered objects. The first object is the gathered centre and everything else is not. In other words, the 'first' object is the foundation of the entire subsequent sequence and yet by itself is not a sequence at all. So, oddly, non-sequence is the foundation of sequence.

Also, since the whole structure of a sequence is relative and positional, each subsequent ordinal relates to all previous objects. For example, 'third' means nothing on its own without 'first' and 'second'. Therefore, subsequent ordinals are less like additions to the first ordinal and more like dimensions or aspects of it. This may sound odd because we so often think in terms of cardinal numbers. But ordinal sequences are different kinds of objects. Here is the philosophical crux of ordinal objects: without previous objects, latter objects would not be 'latter', and the former could not be 'the former' without the latter.

Another fascinating oddity of ordinal sequences is that as each subsequent object becomes larger than the initial object, the initial object becomes retroactively less, relative to the most recent addition. In this way, sequences create a constant disequilibrium in relative position. Each new addition to the series changes the relative position of every other object in the sequence. In other words, each new ordinal object is not genuinely separate but transforms the whole series. Each object in the sequence is not a cardinal 1, 2 or 3, but a dimension or aspect of an immanent order. For example, when we add rocks to a pile, all the rocks below become slightly 'deeper' in the pile than they were.

The *last* object in a sequence is similarly paradoxical. We can always add more to any sequence and change the whole relative series. But if each new ordinal object changes the series, then the series is always in flux. As such, the distinction between the centre and periphery is not static. If we can always add one more object, the centre can invariably become more than it was. This also means that the periphery can always become smaller than it was. Sequences are thus fundamentally unstable.

Another remarkable aspect of ordinal sequences is their fundamental ambiguity. Every object in a series, except the first and the last, is both 'less than' and 'more than' other ordinal objects simultaneously. The second rock is more than the first rock but less than the third rock. There is no totality or sum of all objects in an ordinal sequence that could resolve its ambiguity.

The French neuroscientist Stanislas Dehaene accurately compares early human and animal ordinal counting with how water flows gather into a bucket.[5] People did not count the water flows as isolated individual objects but as transformative dimensions of the relative volume of water. Early humans and animals can compare and contrast the water volume in the bucket with other volumes, as greater or lesser, but they do not count a precise unit of difference in the flows or volumes.

In this sense, ordinality is much more like spiral arms than concentric circles. Spirals have no beginning and no end. They only have 'arms' that are relatively closer to or farther from the centre or periphery. So it is with ordinal sequences. Like water flowing into a bucket, each new spiral arm is not a separate arm but an expansion and transformation of the whole spiral. Like an ordinal sequence, the movement of a spiral also transforms the centre and the periphery simultaneously.

One-to-one coordination

The third feature of ordinal objects is their unique method of coordination. We can coordinate two or more sequences to one another through a process of 'one-to-one' coordination. Imagine we have two series, one of rocks and one of pears. If we lined them up side by side in two rows, we could match one stone to one pear down the series. We would not know 'how many' rocks or pears there were, but we would know if the sequence of stones was greater, smaller or equal to the series of pears at the end of the process. We can see here that the rocks do not 'represent' the pears or vice versa, nor is the coordination activity a strictly 'mental' event. The two sequences remain qualitatively different and related to one another through the concrete kinetic act of physically matching them up side by side.

Particular objects oscillate in cycles of day and night, wake and sleep, eating and defecation, breathing in and out, up and down, front and back, left and right. When the bipedal walking body oscillates its left and right legs, it also sways its left and right arms. Even the earliest human-made tools, such as the biface hand-axe, were made by oscillating tool-strikes back and forth on each side of the axe. Humans raised their arms, struck one rock with another, repeated, flipped the rock, raised their arms again, hit one stone with another again, and repeated. All these oscillating cycles have an ordinal sequence because they alternate one after the other.

Throughout millions of years, humans began to coordinate these sequences of cycles. Bipedalism allowed early humans to coordinate gesture and sound in language and tool-making sequences into repeatable and transmittable techniques.[6]

The invention of tally marks allowed humans to coordinate increasingly diverse objects, even though a tally does not represent anything.[7] Instead, it is a kinetic technique for coordinating two or more ordinal series. More specifically, it creates 'one-to-one coordination' where each object in a first series matches kinetically another in a second series, like the series of rocks and pears. The two series coordinate, 'one-to-one'.

What is the kinetic pattern responsible for these tally marks and coordinations? The hand rises, comes down on a point, marks a line, rises again, and repeats. These hand movements form an ordinal sequence of cycles. Each cycle in the series can coordinate with the cycles in other series such as solar cycles, lunar cycles, planetary cycles, and life and death cycles.

This was indeed an amazing discovery. The simple performance of moving one's hand up and down with a rock on a bone records a sequence of cycles that can be matched one-to-one with the series of cycles for other objects. The French archaeologist André Leroi-Gourhan was right to describe this kinetic performance of making tallies as a rhythm. 'Elusive as they are, these marks denote deliberate repetition and, consequently, rhythm.'[8]

This is what most theories and histories of science and mathematics overlook. They tend to think only about the tally mark itself *as a product* and fail to see how the material processes that produced the mark shape its structure and meaning. Rhythms and sequences are everywhere in nature. The novelty of the human-made ordinal object was in coordinating the rhythms together into composite chains.

Any sequence can coordinate with any other, but since each series is qualitatively unique, so is the coordination. This is evident in the case of coordinating a series of rocks with one of pears. But tally marks are just as singular. Each tally mark and the action that made it is unique. This is true of any inscription. No inscription can happen again in the same way because a second 'duplicate' inscription will have come *after* the *first* and will thus be sequentially distinct. Ordinal objects are, therefore, singular and relational.

I hope the reader can see how this one-to-one coordination is not about resemblance or representation but instead about coordinated kinetic cycles.[9] The ideas of ratio, proportion and correspondence are later inventions, and we should be careful not to project them backwards in history or universalise them. The singularity and indivisibility of ordinal objects make them profoundly different from cardinal numerals. An ordinal object is not a unit that we can subdivide into other units. The position of an object in a sequence cannot change without changing the whole series.[10]

Conclusion

In this short chapter, I have described the purpose and direction of Part II of this book, which is to elaborate a theory of objects built from the history of four main kinds or orders of objects. I began the first section of this four-section history by introducing the broad strokes of the first kind of object that emerged in prehistory: the ordinal object. Instead of offering a formal definition of the ordinal object as the English philosopher Bertrand Russell did,[11] I tried to give a more materialist and process-oriented definition. In particular, I have shown that ordinal objects are not abstract ideas but are the result of centripetal motions, sequential cycles and kinetic acts of one-to-one coordination.

What we need to look at more carefully now, however, are all the historical methods of prehistoric science that created and sustained these kinds of object. This is the goal of the next chapter on the rise of ordinal objects in the prehistoric field.

Notes

1. I do not mean to suggest that there is any progress, evolution, or any superiority of the Western tradition. It just happens to be the tradition that has shaped me and my research.
2. See Tobias Dantzig, *Number: The Language of Science* (New York: Plume, 2007).
3. Georges Ifrah, *From One to Zero: A Universal History of Numbers* (New York: Penguin, 1988); Raymond Wilder, *Evolution of Mathematical Concepts* (Mineola, NY: Dover, 2013); Dantzig, *Number: The Language of Science*; Peter Rudman, *How Mathematics Happened: The First 50,000 Years* (Amherst, NY: Prometheus Books, 2007).
4. See Section II.
5. Stanislas Dehaene, *The Number Sense: How the Mind Creates Mathematics* (Oxford: Oxford University Press, 2011), 18.
6. I discuss these in more depth in Thomas Nail, *Theory of the Image* (Oxford: Oxford University Press, 2018).
7. Maxine Sheets-Johnstone provides one of the most interesting accounts of counting in the origins of bipedal human motion. Unfortunately, her conclusion remains rooted in the human body and not the more primary periodicities of the world itself. My argument is the inverse: it is only because the world is periodic that human periodicity can sync to it. Counting is thus not a projection of human embodiment, but human embodiment is already an expression of material counting and synchrony in the first place. The human body is derived from natural sync, not the other way around. See Maxine Sheets-Johnstone, *The Roots of Thinking* (Philadelphia: Temple University Press, 2010).
8. André Leroi-Gourhan, *Gesture and Speech*, trans. Anna Bostock Berger (Cambridge, MA: MIT Press, 1993), 370.
9. Sheets-Johnstone, *The Roots of Thinking*, 86.

10. See Henri Bergson, *Time and Free Will*, trans. Frank Lubecki Pogson (New York: G. Allen, 1913), 82: if a multiplicity 'implies the possibility of treating any number whatever as a provisional unit which can be added to itself, inversely the units in their turn are true numbers which are as big as we like, but are regarded as provisionally indivisible for the purpose of compounding them with one another'.

11. On ordinality, see Bertrand Russell, *The Principles of Mathematics* (1920) (London: George Allen and Unwin, 1964), introduction.

5

The Prehistoric Object

This chapter looks at the historical emergence of ordinal objects in prehistoric science. My argument here is that these ordinal objects are neither static forms nor mental constructs. They are metastable states created by the three kinetic processes described in the previous chapter.[1]

This chapter shows how these processes are at the heart of three major prehistoric sciences: tool-making, signs and tallies. My aim here is to demonstrate that prehistoric peoples invented many different kinds of ordinal objects through the same centripetal pattern.

Tool-making

Tool-making is one of the oldest human activities. I want to argue here that tools are ordinal objects formed through centripetal patterns of motion. But what is a tool? As I hope to show, a tool is not just a single object, but an ordered series of objects. More specifically, it is a *sequential* chain of objects.

'Tools', as the French archaeologist André Leroi-Gourhan argues, are 'found in invertebrates as much as in human beings and should not be limited exclusively to the artifacts that are our privilege.'[2] In a broad sense, Leroi-Gourhan is right. The mobile body is already a kind of tool. There is no absolute division between a user and a tool, but only a relative difference depending on one's position in the operational sequence. There is no user, though, who is not also a tool of something else.[3] There are only objects and relations woven from flows of matter in various orders.

Ultimately, I do not think we can define a tool in these instrumental terms. Instead, we need to look at the order of the whole field of tool-objects.[4] As I propose defining it here, a tool is a whole ordinal series of objects working together. This means that the materials, the maker, the process of making, the made object and the object's usage are all part of 'the tool'. In this sense, a tool is not one discrete or static object but a metastable circulation of objects.

Some of the earliest tool-making sequences began 3.4 million to 2.5 million years ago.[5] Our ape-like hominid ancestors, Australopithecus, began gathering stones to a central worksite and started making the first stone tools. What was the kinetic structure of this earliest tool-making? First, they had to collect two different kinds of rocks from the East African grasslands' vast periphery. They needed a hard 'hammerstone' to strike with and a softer 'core stone' to strike. The sequential order of the stones was critical. The core stone had to be weaker than the hammerstone.

However, the first object in the ordinal sequence was a firm and flat patch of ground where they held the core stone. From the diverse surrounding terrain, tool-making hominids gathered together at this central location to work. They created a unique sequence of ordinal objects in a particular order with the flat surface on the bottom, the softer rock on top, and the harder rock on top of that. By grasping the hammerstone, the hominid body was at the end of the sequence. Tool-making requires the centripetal collection of all four of these objects in a particular series.

Once hominids gathered these objects together in this sequence, they began to cycle through the order repeatedly. Specifically, they raised their arm with a hammerstone, lowered it to strike, and repeated. The arm is a tool of the body, and a hammerstone is a tool of the arm. The kinetic structure of tool-making is an iterated sequence of cycles. The ground, core, hammer, arm and body all become entrained in the process. Stones, bodies and brains all entered into an ordinal sequence where each transformed and worked on the other. The body shaped the rocks, and the stones' weight shaped the body's muscles. The specific series of rock types and bodily movements also shaped the brain, which learned and refined the tool-making method.

This ordinal 'knapping cycle' allowed human cognition and tool-making to co-evolve side by side. Each new tool is only one aspect of an entire kinetic field. In this way, tool-making and technology transform human minds and bodies simultaneously as these minds and bodies transform technology.

Entrainment

Entrainment occurs when several motions merge or mix into a single flow. This is what happens in the centripetal ordering of an ordinal sequence. For example, the circadian rhythms of an organism entrain with the progression of night and day. Sediment in rivers entrains with the flow of water from higher elevations to lower ones. Pendulums moving back and forth on the same wall of a house will also entrain their motions back and forth. The phenomenon of entrainment describes how divergent and

heterogeneous processes gather together into a sequentially ordered feedback loop.

Ordinal objects entrain into sequences of cycles, and epicycles transform one another. This is also what Leroi-Gourhan calls a *chaîne opératoire*
or 'operational sequence'. In an ordinal field, the human body becomes
rhythmically entrained with the material world like a series of pendulums
slowing syncing with one another. An operational or ordinal sequence is
not created by applying a top-down mental template on to a passive bit of
matter, but instead emerges through 'functional plasticity' where the entire
series responds to and alters itself.[6] The principles of ordinal sequence are
'derived from the laws governing matter and cannot, by that token, be
regarded as human attributes except to a very limited extent'.[7]

For example, the Acheulean biface, a pear-shaped, sharpened, stone
hand-axe, did not originate in the hominid brain. It did not begin as a
blueprint or idea merely applied to the passive stone material, like Plato's
divine craftsman. Instead, the biface hand-axe is only one object in a whole
field of sequentially coordinated objects that all played a unique role in carefully knapping a rock on both sides without breaking it. Through knapping, the body and brain became more rhythmically entrained with the
stones' shape and hardness. Harder stones entrained with softer stones with
more cuts. By changing the sequence of hammering cycles, hominids could
produce new objects.[8]

However, the movement of hammering, sawing and chipping had
first to become rhythmically entrained with the physical properties of the
various stones. Hitting a rock too hard could break it. Not hitting it hard
enough might not chip it. Therefore, tool-making required sequences,
cycles, iterations and the entrainment of a whole field of objects. Tool-
making is much more like improvisation than a blueprint.

What comes first is not the form or mental blueprint of the tool-object,
but rather the centripetal motion of gathering rocks from the periphery
and entraining them in a sequence. What the hominid body and brain can
do co-evolved with the ways certain types of stones could be broken and
used. The agency of objects and the agency of the hominid body are both
at work in tool-making.

Coordinating tools

Every tool is an ordinal series that one can coordinate with other ordinal
series. For example, several coordinated series shape the Acheulean biface.
The first series is the centripetal gathering and sequential ordering, from
bottom to top: ground, core and a hammerstone. The second series is the
sequential cycle of human arm motions: raise, strike, repeat, rotate the core

stone, repeat. The third series then coordinates the first two. Instead of rais-
ing the hammerstone, the arm extends the biface stone to strike an animal,
branch or nut. We can now see that the whole tool-using sequence is an
iteration or continuation of the tool-making series by other means.

The idea that tools are useful instruments for a pre-given end is backwards. How can we know what an object can do before we use it? We
have to start by using the object in different ways. Then maybe we will
like one of those ways and keep doing it. As we do it more, we experiment
with variations to see if what we wanted gets better or worse. In this way,
we learn to use tools by responding back and forth with objects' unique
agencies. There is nothing mysterious about this 'bottom-up' explanation.
It is an improvisation with the sequence of objects.[9]

In the case of the biface hand-axe, hominids improvised based on coordination with the knapping cycle. They used the biface like a hammerstone
to strike other objects to see what they could make. The results were all
kinds of differently shaped objects that our hominid ancestors liked and
found a use for. The benefit came at the end of the process, through improvisation and iteration, and finally took on a relatively stable form matched
to its selected use. We call these refined objects 'bowls', 'wooden stakes',
'clubs' and 'huts'.

Humans used each tool-series to make more tools, and each new tool
changed the whole field of what tools could do, relative to the new tool.
In this way, tool-making works like the ordinal series of objects I described
in the previous chapter. Eventually, Neolithic humans started making tools
that increasingly did more of the tool-making activity. For example, the
hand-operated milling machine incorporates the 'arm-raising, striking,
repeating' cycle into the machine's movement. In this case, the human arm
takes on a new cycle as the motor for the device. The history of technology
is diverse, but the kinetic coordination of sequences is at its core.

Early humans coordinated their sequences of tool-making through
what we might call 'centripetal transmission'. That is, they made their
tools together in particular places. Not only did early humans select and
move objects from the periphery to a central region to assemble them in
a sequence, but human bodies themselves also had to gather in concentrations. In this way, humans learned, improvised and coordinated one
another's arrangements. Humans learned technology socially and through
this same structure of one-to-one coordination. As an older human raised
their hand with a hammerstone, a younger one raised its hand; as the older
one struck with a particular force, so did the younger one, until each detail
of the sequence was iterated one-to-one.[10]

We can see Brazil's bearded capuchin monkeys doing the same thing
today. They harvest nuts, wait until they dry out, place them in pitted

stone holes, and then hit them with a hard rock until they break open. If they don't do this in this precise order, the process does not work. The tool-sequence fails to produce food. The younger monkeys gather at the same rocks around the older ones and practise the same one-to-one coordination. It takes them years to learn it, but eventually they do, and share the technique with others. In this way, the tool-sequence reproduces and transmits itself through history via the body's biological medium.

Signs

Signs and various tool-made marks are a critical part of prehistoric science and knowledge, but have been under-theorised. I want to show here that these signs are also ordinal objects produced through centripetal patterns of motion. But what is a sign? A sign, as I hope to show, is an ordered series of centripetally gathered marks.

To our knowledge, the first non-figurative signs emerged during the Lower and Middle Palaeolithic periods (300,000–200,000 BCE). This is long before figurative art appeared in the Upper Palaeolithic (35,000–10,000 BCE). Given this, I am surprised that relatively few scholars have paid attention to the vast diversity of these signs made by early humans. These markings include points, lines, arrows, squares, circles, triangles, spirals, meanders, chevrons, V, Y, M and P shapes, and many others.[11]

Some of the earliest signs, such as V, M and parallel lines, were engraved on an Acheulean-era rib bone from Pech de L'Azé, France. These same signs persisted into the Neolithic period, hundreds of thousands of years later.[12] Given the relatively small number of these different signs (about 32), their geographical near-ubiquity from Africa to Europe, and their persistence from the Acheulean to the Neolithic, these markings certainly played an essential role in early human knowledge.[13]

Unfortunately, these signs have been ignored or dismissed as doodles or decorations. To describe these markings as 'doodles' or 'decorations' implies that they are meaningless fragments unworthy of explanation. But why was this set of signs used and not others? How and why were they selected, gathered and reproduced for hundreds of thousands of years? Calling them doodles does not answer these questions but dismisses them.

Furthermore, if these signs were doodles or decorations, why don't we see a more extensive range of geographical and temporal variations in them over time? Thirty-two different signs is a relatively small number. Random doodles and meaningless decoration almost certainly would be much more multiple and diverse.

However, what does seem clear is that these signs are not a prehistoric alphabet, a written language, or pictographic representations. Early humans likely had much more extensive vocabularies than thirty-two pictographs could represent. Also, most of the signs are not even remotely figurative, so they are probably not pictographic. These signs are much more abstract marks that precede written language and pictograms. As Leroi-Gourhan rightly notes, 'there is one point of which we may be absolutely sure, it is that graphism did not begin with naive representations of reality but with abstraction'.[14]

Of signs and sequences

I want to argue that these early human signs are not pictographic but rather 'kinographic'. By this, I mean that humans did not intend them to 'look like' something else but to leave a trace of a sequence of actions. In other words, the sign is not 'about' something else, but more like a trail left by a rhythmic and ordered sequence of bodily motion. Just as a mental blueprint did not precede the first tools, it did not precede these earliest signs. The shape and meaning of the sign emerged simultaneously as two aspects of the same rhythmic performance.[15]

Each sign has its rhythm depending on how many strokes it has and how long each stroke is. By treating prehistoric signs as static objects, we miss the movements, sounds and sequences embedded in them. My argument here is that these signs were marks that were coordinated one-to-one with other movements at the same time. These might have included a story, a song, a description or a dance. The sequence of strokes for making each mark and the sequence of marks themselves might have worked like a mnemonic device for another performance. Early humans might have coordinated each stroke to one part of a story. In this way, the signs were not about representing the story but were ordinal objects coordinated to the story.[16]

It is no coincidence that humans started marking signs at the same time as they began making tools. Both techniques follow the same centripetal pattern of motion. Humans gathered hard rocks and other materials from the periphery and accumulated them in central caves. Then, they repeated this for thousands of years. The same pattern of bodily motion that made tools also made signs. The arm raised a hard rock, struck or scraped a softer rock with it, and then repeated the sequence. In the case of signs, however, the result of this ordinal sequence was a series of marks and not another physical tool that could make more tools. Each sign is a sequence of gestures whose force, duration, frequency and angle were coordinated one-to-one with some other sequence.

What were these other sequences of motion? There is no way to know for sure, but they might have been of trees moving back and forth in the wind, animal sounds, the expansion and contraction of the lungs, the echo of sounds in the cave, or thunder.

The South African archaeologist David Lewis-Williams has shown that even in the complete darkness of a cave, the eye can create 'entoptic images'. These images inside the eye match the grids, lines, zigzags, dots, spirals and curves of prehistoric signs.[17] However, these entoptic images do not account for all the thirty-two different signs in Ice Age caves.

What seems most likely to me is that these entoptic images were only part of the visual inspiration for prehistoric signs. Nature has order, sequence and symmetry that were also an inspiration for these more abstract signs. The critical point in my view, though, is that the marks on the cave walls were not representations of entoptic images or natural patterns, but were rhythms made by a human arm in coordination with a series of other motions.

The person who wrote these signs did not first imagine a perfect geometrical shape in their mind and then try to copy it on the cave wall. Nor did they randomly scratch away at the wall without any order. Instead, they started by improvising with the materials of various rocks and pigments to see what they could do when used in different series. Then, once they found a sequence they liked, they coordinated it with other progressions. In this way, they coordinated two sequences simultaneously in the same rhythmic act of inscription.

Bipedalism was so revolutionary because it freed up the hominid hand and mouth at the same time. Just as the freer use of hands allowed hominids to make tools and signs, their free mouths allowed them to make all kinds of new noises. Naturally, they increasingly coordinated these two new kinds of sequences in one-to-one relations.[18] If someone or something made a particular sound simultaneously with a specific gesture or mark, then the sound and motion were coordinated one-to-one. Over time an entire sequence of sounds could be coordinated to just one gesture or sign, and vice versa. This is what I believe we see in these prehistoric signs. We do not see representations but kinetic coordinations of ordinal sequences.

These prehistoric signs are not attempts at universal geometrical shapes. To think so would be anachronistic. If there is anything like a 'prehistoric geometry' in these signs, it has nothing to do with perfect, unchanging, Pythagorean forms. Each sign is singular, mutable and performative. Each time one draws the sign, it iterates and changes the shape slightly. It might have even summoned again, as if for the first time, its coordinated partner. These signs' geometry was bound to the rhythmic sequence of *how* and *where* one drew it.

The 'point' sign, for example, has a short rhythmic burst. The line has a longer duration, and the cruciform (cross) has two long rhythmic motions. There is a tendency among archaeologists to think of these signs as complete products. Yet in all likelihood, the signs can only be understood in their inscription's performative context. This context, however, is lost to us.

However, whatever their context, it was most certainly a kinetic one that bound the graphic and sonic sequences together. Often, early humans would iterate the same sign several times in a row or group. We can see this in the Spanish caves of *Grotta dell'Addaura* and *Riparo di Za Minica*, where open angles, lines, cupule (holes) or dots iterate in stacks, parallels, or in an overlapping sequence.[19] Here, the ordinal and rhythmic structure is striking. These signs were not letters in an alphabet or representations of other things, but objects in their own right. They have rhythms and sequences that humans worked with and coordinated.

Matter and memory

Prehistoric signs are traces of coordination between a body, a tool, a gesture and a sound. Signs do not represent the world but remind the prehistoric viewer of specific rhythmic coordinations between a series of gestures and sounds. Walls or mobile objects with signs on them are material memories. They remind the viewer of a sequence of something, then something, then something else, that happened here, then here, then there.

This is not an *externalisation* of human memory on to the world but a coordination of one series of ordinal objects. The idea that signs are externalisations of human mental images on to the world of passive objects ignores the fact that non-human objects already have their own material memories. Every object retains a material trace of its history in its physical structure. This is no different from human brains and bodies, which also possess such physical traces. If this is the case, then sign sequences are not representations of human memories but coordinations between two material memories.

Tallies

Tallies are signs, but the most abstract of all signs, since they were used by the earliest humans to measure the duration and extension of another ordinal series. Given the diversity of signs above, they probably coordinated all kinds of different sequences. Tallies, however, were not nearly as diverse. They were variations of lines. In this section, I want to show that, like tools and signs, tallies are ordinal objects shaped by the same pattern of centripetal motion and one-to-one coordination.

Tallies are pretty clearly not figurative or representational, but that does not mean that they are entirely abstract quantities. They are a series whose purpose is to show 'how much' or 'how long' another series is by degrees of 'more or less'. This is why tally marks are more homogeneous in form than the more geometric-looking signs described above. The more homogeneous the mark, the less the writer emphasises the mark's individuality, and the more its spatiotemporal differences are accentuated. A beautiful example of this is the famous 30,000-year-old 'wolf bone', which some early human marked with 55 relatively identical parallel lines ordered into small groups. The sheer quantity of nearly identical repetitions suggests that the writer coordinated each line in a one-to-one relationship with only one other object. The iteration then works to express the amount and duration of the quality of this other object.

The moon

The moon was maybe the earliest object to be measured. A small trace of this idea remains in the etymology of the word 'moon', which comes from the Latin word *mensis*, meaning 'month', and *mensura*, which has the same Proto-Indo-European root *-me*, meaning measurement. The moon's sequential changes made possible the earliest recorded measures longer than the length of a day.

Nature moves in cycles, but the moon is one of the earliest and most dramatically visible cycles. The sun is more binary. With the brief exceptions of sunrise and sunset, it is either full of light or gone. The moon, however, is visible day and night and waxes and wanes in regular sequence. The moon's full light is only one position in a shifting ordinal series, not a cardinal sum. Eventually, in history, the sun would take over as the great image of all objects, but the moon was still the tremendous ordinal object of becoming in prehistory.

Like the gathering of notches on bone, the light of the moon follows a centripetal pattern of movement. The moon gathers the sun's light from the periphery and reflects it off its central disc. The moon reflects more or less light depending on the relative position of the earth. As the moon waxes, it reflects more, and as it wanes, it reflects less.

The sun's rising and setting cycle provides a relatively stable background to measure the more dynamic fluctuations of the moon over the month. There are always more or fewer lunar days relative to the lunar month, while the solar day begins and ends again every time.

Prehistoric tallies also counted the new moon's duration, which is the darkest phase of the moon. This is a fascinating idea because it is essentially counting what is not visible. In this way, the new moon's ordinal measurement was the precursor to 'zero'. The dark moon is the empty surface that will fill with light like the flat surface required by the tool sequence.

The new moon is the first and paradoxical object of the ordinal series. It is the dark centripetal stage or empty place that slowly fills with light until it reaches the limit of its sequence at the full moon.

The Blanchard bone is one of the oldest lunar bone tallies, made between 25,000 and 32,000 years ago. Under microscopic analysis, this small, 4-inch bone shard reveals that someone made a series of sequential notches with slight curves, corresponding to the waxing and waning 'turns' of the moon. Each mark was made one at a time over a long period. Most scholars either ignored these marks or considered them merely random 'decorations' before the American archaeologist Alexander Marshack used microscopic analysis to show that the tallies were grouped and curved according to the moon's phases.

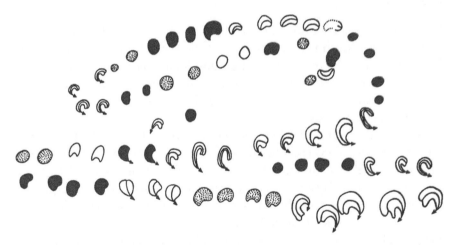

Figure 5.1 Engraved marks on the Blanchard bone. From Alexander Marshack, *The Roots of Civilization: The Cognitive Beginnings of Man's First Art, Symbol and Notation* (Mount Kisco, NY: Moyer Bell, 1991), 48.

Marshack described several lunar tally bones with similar kinds of markings. For example, the Lebombo bone is a baboon fibula that contains twenty-nine notches, the number of days in the lunar month, made with different cutting edges over a series of time.[20] Someone carved it between 44,200 and 43,000 years ago. Another two tally bones, the Ishango bone and the 'Second' bone found alongside it, are likely lunar tallies as well. Both are around 20,000 years old. The Ishango bone has two columns of sixty temporally sequential tallies each (very close to lunar months), and the Second bone has twenty-nine.[21]

Ritual

Tally marks are ordinal objects that one uses to coordinate any other cyclical sequence. Early humans used tallies to record lunar cycles, menstrual cycles and seasonal cycles. Tally marks recorded the frequency of one kind of coordinated object over time. In this way, they were singular coordinations and not universal. One can always add one more to a sequence.

The ordinal structure of tallies also made it possible to record ritual participation. Natural cycles repeat relatively regularly, but social cycles are much more irregular. For example, archaeologists have discovered tally marks notched in prehistoric bone flutes. According to these archaeologists, the notches seem to have been incised at various times and with different tools. The incisions have no apparent relation to the instrument's construction or decoration, and some of the notches were cut after the holes were made. Therefore, archaeologists suggest that people probably made these ordinal notches in coordination with the *instrument's ritual use*. The tallies start widely spaced then become more narrowly spaced as they approach the end of the instrument. This may have indicated that the inscriber increasingly ran out of room to keep tally. Furthermore, it is likely that more than one person played the flute and that each time it was used by another person, they 'made their mark' on it.

Early humans transformed the flute into a ritual object by coordinating its use to the ritual events where people centripetally gathered. Early humans gathered marks on to the flute's body, just as they gathered marks around other marks, just as they gathered people near one another in the caves. We can see here that a centripetal motion dominated the structure of ritual.

There are at least three coordinated ordinal series at work here: the playing of the flute, the person making the mark, and the number of marks. Ritual tallies coordinate all three ordinal series on the centripetal surface of the flute.

There is also an incredible number of Palaeolithic and Neolithic *batons* or bone pieces with holes. These *batons* were worn as necklaces and covered with tally marks made by different edges over a long time. Some of these *batons*, Marshack suggests, were coordinated directly to lunar cycles. People might have coordinated others to other 'processes, sequences, and periodicities of female, sky, and season', or connected to larger 'time-factored' ritual, ceremonial events.[22]

It is no coincidence that the structure of oral storytelling and myth is deeply ordinal. Stories are sequences of sounds and gestures that are repeatedly told and are slightly different each time. Ritual tallies served as *aides-mémoires* in the form of a portable necklace because the tallies and stories are both ordinal series. The markings that remind the storyteller about what happens next in the story do not need to be figurative to provoke memory. It is precisely the ordinal structure of the notches that is most appropriate to the ordinal form required by the mythological narrative.[23]

Each repetition of an oral story or a ritual does not add a second or third time to the first, but rather iterates the first time exponentially. In this way, prehistoric myth does not move like a circle but like an iterative spiral. Each new tally and tale changes the relative meaning and position of the previous iterations.

Ordinal objects do not form a whole but a dynamic, meandering and labyrinthine process. Ritual follows a centripetal motion as it gathers people to a centre for the event and creates a sequence of events. Tallies are the coordinated traces of the ordinal structure of ritual.[24]

Conclusion

In this chapter, I have tried to show how the first kinds of scientific objects emerged through the prehistoric practices of tool-making, signs and tallies. I hope the reader has also seen how the three primary features of ordinal objects that I described in the previous chapter emerged through these concrete historical practices.

Furthermore, I also hope that it is clear from the evidence above that there were no cardinal kinds of objects, such as numerals, in prehistory. We should not forget that the prehistoric answer to the question, 'How many moons cycle in a month?' is not '29 and a half', but rather 'this many'. One of this book's main aims is to show that objects are not universal forms, but emergent patterns of motion.

The materiality and mobility of prehistoric objects also play a crucial role in the emergence of these movement patterns. For example, the sequence of rocks in the knapping cycle is not a construct of human brains, but of rocks' real capacity to be harder and softer. Similarly, the idea of counting unseen objects was not only human; the moon's particular motion and agency were part of it.

I also hope this chapter has clarified that the difference between ordinal and cardinal objects is not a formal or logical one but a material and historical one. However, if cardinal objects are not pre-given features of reality, the next question is, 'How did they ever emerge from ordinal objects?' If the emergence of new kinds of objects is not merely a human construct, what kinds of kinetic patterns were required to create and sustain it? These are the questions of our next section on cardinal objects.

Notes

1. 'Centripetal movement', 'sequential cycles' and 'one-to-one coordinations'.
2. André Leroi-Gourhan, *Gesture and Speech*, trans. Anna Bostock Berger (Cambridge, MA: MIT Press, 1993), 237.
3. Graham Harman, *Tool-being: Heidegger and the Metaphysics of Objects* (New York: Open Court, 2011). I do not agree that the 'reality' of the tool always withdraws from all relation. I do not think conversely that relation precedes *relata* either. Both occur at the same time in the same field of motion. Furthermore, matter, for me, is not an ontological determination but a specifically historical ontological determination. See Thomas Nail, *Being and Motion* (Oxford: Oxford University Press, 2018), on the issue of history, method and process materialism.
4. See also Thomas Nail, *Theory of the Image* (Oxford: Oxford University Press, 2019), for a theory of 'functional kinesthetics' vis-à-vis so-called 'instrumentality' in art.
5. For a long time, no known stone tools were associated with *A. afarensis*, and paleoanthropologists commonly thought that stone artefacts only dated back to about 2.5 million years ago. However, a 2010 study suggests that the hominin species ate meat by carving animal carcasses with stone implements. This finding pushes back the earliest known use of stone tools among hominins to about 3.4 million years ago. T. L. Kivell and D. Schmitt, 'Independent Evolution of Knuckle-walking in African Apes Shows that Humans did not Evolve from a Knuckle-walking Ancestor', *Proceedings of the National Academy of Sciences of the United States of America*, 106.34 (2009), 14241–6.
6. Leroi-Gourhan, *Gesture and Speech*, 301.
7. Leroi-Gourhan, *Gesture and Speech*, 303.
8. For a full account with numerous examples, see Gary Tomlinson, *A Million Years of Music: The Emergence of Human Modernity* (New York: Zone Books, 2018), 309: 'One of the operating techniques of human beings of the earliest stages has been the application of rhythmic percussive movements repeated over prolonged periods. Indeed that is the only operation that marked the attainment of human status by the Australanthropians, whose only surviving traces are choppers made of splintered pebbles and polyhedral spherical objects produced by prolonged hammering. Manufacturing techniques developed from the beginning in a rhythmic setting – at once muscular, visual, and auditive – born of the repetition of impact-making gestures.'
9. Karl Marx, *Grundrisse: Foundations of the Critique of Political Economy*, trans. Martin Nicolaus (New York: Penguin, 2012), 81–9.
10. Tomlinson, *A Million Years of Music*, 33, 'group musicking'.
11. For a full typology of these signs, see Genevieve Von Petzinger, *The First Signs: Unlocking the Mysteries of the World's Oldest Symbols* (New York: Atria Press, 2017).
12. Marija Gimbutas, *The Living Goddess* (Berkeley: University of California Press, 2005), 43.

13. For important qualifications to the ubiquity, scope and duration of these symbols, see Von Petzinger, *First Signs*. Not all signs are in every region, and not all signs are as common as others. Some signs are clustered in certain periods and areas, and so on.
14. Leroi-Gourhan, *Gesture and Speech*, 188.
15. Leroi-Gourhan, *Gesture and Speech*, 309.
16. Based on the development of ordinality in tool-making and the emergence of tools it is thus not unlikely that graphism preceded even the oldest European marks, and goes all the way back to African cave and mobiary art. However, research is still need to confirm the prevalence of graphism this far back.
17. See David Lewis-Williams, *The Mind in the Cave: Consciousness and the Origins of Art* (London: Thames and Hudson, 2002). See also Von Petzinger, *First Signs*.
18. I deal with the relation of speech and language more closely in *Being and Motion*.
19. See Von Petzinger, *First Signs*, 156–7.
20. Peter B. Beaumont, 'Border Cave – A Progress Report', *South African Journal of Science*, 69 (1973), 41–6.
21. Marshack discusses the discovery of what he refers to as a 'sister bone' with similar markings: 'from the same site, the same period, and the same level, there existed a bone of the same bird, engraved in exactly the same style . . . clearly what we have is the same style and technique as on the first bone'. Alexander Marshack, *The Roots of Civilization: The Cognitive Beginnings of Man's First Art, Symbol and Notation* (Mount Kisco, NY: Moyer Bell, 1991), 159.
22. Marshack, *The Roots of Civilization*, 283.
23. Claude Lévi-Strauss, *Myth and Meaning* (London: Routledge, 2014), 39–40.
24. Maurice Bloch, 'Why Religion is Nothing Special but is Central', *Philosophical Transactions of the Royal Society*, B 363 (2008), 2055–61. This essay is reprinted in Bloch, *In and Out of Each Other's Bodies: Theory of Mind, Evolution, Truth, and the Nature of the Social* (London: Paradigm, 2012). See also Michael Silverstein, 'Metapragmatic Discourse and Metapragmatic Function', in John A. Lucy (ed.), *Reflexive Language: Reported Speech and Metapragmatics* (Cambridge: Cambridge University Press, 1993), 33–58 (36); for systematicity and 'configuration' of indexes, see pp. 43, 47–8, 54–5; for ritual, see p. 48. For a further developed view of indexical systematicity and ritual metapragmatics, see Michael Silverstein, 'Indexical Order and the Dialectics of Sociolinguistic Life', *Language and Communication*, 23 (2003), 193–229, esp. 201–4.

II THE CARDINAL OBJECT

6

The Centrifugal Object

Cardinal objects are the ones we think about most often. We typically count things using 'whole numbers' and tend to think of the things around us as discrete self-identical unities. This is our default view of objects, so much so that we assume that humans have always thought about objects this way. However, for 99 per cent of human history this was not the case, as I tried to show in the previous chapter. There is nothing universal or necessary about thinking of things as cardinal objects or whole units. They are a historical invention.

What is so fascinating to me about cardinal objects is how different they are from ordinal objects. Ordinal objects feel very intuitive in the way that kindergarten maths is. Cardinal objects, however, are much more abstract. They are objects that we treat *as if* they had no qualities. This allows us to create enormous and relatively homogeneous kinds of objects in a wide array of contexts. There is no disputing the power of cardinal objects to order the world in apparently firm and definitive ways.

But how did something so abstract emerge from a world of such singular and ordinal differences? This is the subject of Part II of this book. In this first short chapter on cardinal objects, I want to describe their more general structure and pattern of motion, just as I did for ordinal objects in section I. Afterwards, I will try to show more concretely and historically how this new kind of object emerged throughout ancient history from around 5000 BCE to 500 CE. Specifically, I look closely at how numerals, geometric measurement, accounting and logic each helped create this new field of cardinal objects.

Cardinal objects emerge through very different patterns of motion than ordinal objects. Instead of gathering things together into sequences, cardinal objects treat groups of collected things as single homogeneous units. Humans then use these central units to reorganise and measure other objects.

However, cardinal objects do not replace ordinal objects, but rather capture, transform and redistribute them in new ways. Historically and kinetically, cardinal objects came from ordinal series. Eventually, however, ancient peoples came to use cardinal objects *as if* they were the source and foundation of ordinal objects. This is another intriguing aspect of cardinal objects that I try to explain in the following chapters. How did these new kinds of abstract objects restructure the world? What about this new world made it look as though it was created by these objects and not the other way around? It is a genuinely incredible and baffling moment in history when a product looks as though it produced itself.

What is a cardinal object?

Georg Cantor defined a cardinal number as an 'aggregate' of a series of elements 'into a whole'.[1] This is a formal definition, but there is also a more interesting story behind it. When did humans start treating things as homogeneous units that were unchanged by their relative position and context? How was it possible to act as though certain things had no qualities at all?

I believe that, however it happened, it happened in ancient Mesopotamia around 3500 BCE. This is where humans first invented *numerals*. A numeral is a cardinal object such as the Arabic numerals we use today, 1, 2, 3 etc. Numerals are not the same as 'number', which I defined more generally as any magnitude.

This is an important distinction because of the still dominant belief in mathematics, stretching back to Pythagoras and Plato, that number is identical with a cardinal number.[2] 'Mathematics is the queen of the sciences', Carl Gauss wrote in 1856, 'and number theory is the queen of mathematics.'[3] If number theory implicitly assumes, as it does for Gauss, that numbers are fundamentally cardinal, then cardinal numbers are also at the heart of mathematics and the sciences. This tradition also holds that numbers are ideal entities. They are real cardinal objects independent of material reality.

This is a perfect example of where cardinal objects have had the fascinating effect of making the world look, at least to some people, as though it is the product of cardinal objects and not vice versa. Ultimately, this is a metaphysical idea that can neither be proven nor disproven. Instead, I would like to look at the material and historical conditions that made it possible for some ancient human beings to see and live in the world as if cardinal objects created it in their image. What were the patterns of motion and structure of these cardinal objects that made such metaphysical ideas seem so plausible to some?

The emergence of cardinal objects was so dramatic that many science historians equate the birth of 'science' and 'reason' with the advent of cardinal objects, to be precise their later formulation in classical Greece. In the following chapters, I would like to argue a very different thesis. Cardinality is not a purely mathematical idea but rather a material and historical pattern of motion that organises things into abstract wholes, categories and units. In short, I argue that the most abstract idea in the history of Western science is a thoroughly material, historical and practical invention. Cardinal objects are abstract, but an abstraction, at least in my view, is never immaterial, ahistorical, or merely the product of human convention. A whole host of material-kinetic processes are involved in the creation of abstraction.

The following chapters aim to show how these processes created and sustained the ancient field of cardinal objects. In brief, cardinal objects create a model for reorganising, measuring and counting other objects as 'wholes'. In particular, this chapter argues that cardinal objects emerged and propagated in the ancient world through four primary processes: centrifugal motion, concrete cardinality, abstract cardinality and one-to-many coordination. Here, as in Chapter 4, I offer only the very broadest strokes of these processes and how they work. In the next two chapters, though, I describe how these processes spread in various ancient scientific practices. Before getting into history, let's now look at each of these four processes.

Centrifugal motion

By centrifugal motion, I mean a radial movement pattern that emanates from a central point outwards in all directions, like light from the sun. But from where does the central point originate? Centrifugal motion assumes that a group of objects has already gathered into a central point. In other words, centrifugal motion requires an initial *centripetal* motion and thus an ordinal series of objects. Only once a sequence of objects has been gathered is it possible to count this sequence as a closed or 'cardinal' whole. Without an ongoing series of objects to be aggregated, there are no aggregates.

Suppose cardinal objects are a 'collection into a whole . . . of definite and separate objects [elements]', as Cantor says. In that case, cardinal objects require the existence of distinct ordinal objects.[4]

It makes no sense to say that cardinal objects emerged first, from a material and historical perspective. If cardinal objects came first, what did they collect and count as 'one'? Another way to think about cardinal objects is as modified ordinal objects. Once someone gathers enough ordinal objects, one can begin to take on a special status distinct from the others. This remarkable object then begins to function as a model for the others in the group. The other objects are no longer singular but become 'parts' of a greater and more primary whole. As parts, their new order appears to be dependent on the whole. Without the whole, the parts would not be parts.

Once someone designates an object in the ordinal series as a 'cardinal' object, they can use it as a tool to measure, unify and categorise a variety of heterogeneous objects *as if* they were parts of a pre-existing whole. This basic structure of cardinality occurs in all the sciences of the ancient world to varying degrees. All acts of unifying heterogeneous objects into unities are acts of cardinality. Abstract concepts such as right triangles, natural elements (fire, water, earth, air, etc.) or proportionality are all cardinal objects that use wholes to order and define their parts. Ordinal objects, such as tally marks, signs or tools, are not unities in this sense but series of singularities.

This chapter and the next two argue that cardinal objects are created and maintained by ongoing centrifugal movement patterns. In the following two chapters, I will describe the specific historical processes that first generated these objects. In this chapter, though, I want to clarify the two main stages of how this pattern of motion works. Cardinal objects do not emerge suddenly and fully formed but appear in stages of increasing abstraction.

Concrete cardinality

The first stage of the formation of cardinal objects is what I call 'concrete' cardinal objects. Therefore, there are two kinds of cardinal objects that we should not confuse: concrete and abstract. Concrete cardinality is the unification of ordinal objects into a singular cardinal object. Concrete cardinal objects are not abstract because they are still certain *kinds* of groups. For example, a covey of pheasants or a gaggle of geese is a specific kind of unified group. A concrete cardinal object is still a *qualified* quantity.

Cardinal objects began as concrete and became increasingly abstract over time. To help clarify the process, I have divided it into four basic moves. Three of these moves are changes in concrete cardinals, but the fourth one is abstract. By the term 'abstract' here, I mean that people finally treated objects as quantities without any qualities. These steps are crucial because I hope they will help the reader see them at work in the historical chapters that follow.

Division

The first step in creating cardinal objects is marking an arbitrary and firm division between the centre and the periphery. The division divides accumulated objects from others. Ordinal series are always moving and changing their relative positions in continual disequilibrium, like the ceaseless cycle of life, death and rebirth. However, cardinal objects have cut the sequence off at some point so as to be able to count collected objects as 'one' thing.

This results in much starker divisions. For example, in ancient cardinal societies, some plants were deemed healthy to consume (grains), and others not. Some animals could be domesticated, and others (wild ones) could not. As ancient societies gathered more and more ordinal series (of tools, tallies, signs, people and animals) in centralised cities, the contrast between city and wilderness became increasingly distinct. One of the earliest examples of this was the grouping of several agricultural series into a whole 'phase' such as 'a growing season'. People also divided the continuous transformation of the moon into four 'phases'.

Hierarchy

The second step in creating cardinal objects is using the sharp difference between inside and outside to rank objects as more or less superior. Collected objects towards the centre are hierarchically superior to those on the periphery.

Historically, this second step began when Neolithic societies began concentrating knowledge in a select group, elder or chieftain figure. Around this time, rituals and labour started to become more specialised and exclusionary. The emergence of priests and chiefs slowly consolidated the right to initiate prayer, planting and ritual. Neolithic humans introduced the division of labour, although it was by no means universal yet. Humans increasingly concentrated specialised knowledge and skills in privileged places and people. With the rise of urbanisation, the prominent central place became the temple. The darkened interior of the temple functioned as a unique object that was superior to the illuminated outside world.

The late Neolithic distinction between dark and light, destructive and creative, invisible and visible became intensified with urbanisation. Over thousands of years, these distinctions became hierarchical divisions between good and evil and between the superior knowledge of the invisible over the inferior knowledge of the visible. The rise of urban agriculture increasingly privileged the 'whole phase' of the seasons and solstices over the monthly lunar sequences' singularity. The rise of these hierarchical objects was still very *concrete* since it privileged some qualitative object over others.

Reorganisation

The third step in creating cardinal objects followed from the second. Once a single object or group of objects was systematically privileged at the centre of society, it was then used as a model to reorder the periphery. This is what happened when people accumulated grain into a single central silo in late Neolithic villages. Authorities at the central silo had to measure out units of grain to villagers evenly. Without a central stockpile of grain, spe-cialised accountants, standardised units and grain tokens would have been unnecessary, just as they had been for thousands of years before.

In this way, a single cardinal object became the basis upon which all the 'parts' became dependent. People measured out life in homogeneous units of grain.

Abstract cardinality

The fourth step in creating cardinal objects was to treat this new homo-geneous unit as a generic quantity for more than just grain. This final step is not intuitive and even verges on the absurd. How can a unit of grain be used as a universal unit of measure? In what sense are one apple, one cup of grain and one ox all 'one', if they have nothing in common?

Here is where it helps to think about things kinetically. Imagine that each time an event occurred, such as an illuminated moon, instead of draw-ing a unique tally for each one, we drew a tally mark in the same place. At a certain point in the sequence, imagine that we stopped and said all those events are 'one' thing, like a 'moon phase' or 'a season'. A cardinal object is like a process that completes only one cycle in the time it takes many other sub-cycles to cycle. It counts everything that happened while it was cycling as 'part' of its process. It is like a cycle that extends or envelops several sub-cycles.

If the cardinal object 'one' can refer to 'one of anything', something extraordinary follows this abstract experience. If we consider 'one' on its own as 'one of anything in general', it follows that we are treating this 'one' as something without any concrete particularity or quality. In this way, it is as if this general 'one' pre-exists and remains unchanged by any of the par-ticular objects it counts. This is a moment of real abstraction in which one retroactively treats the whole concrete process that led up to the abstract cardinal as a product of the cardinal object.

At the extreme of this process, the one even seems to have *created* the concrete parts that it counts. This is the final and fanatical inversion that rose to historical dominance with Pythagoras and Plato. It completes the four steps of cardinal object formation. Let's look now at how its structure of coordination differs from that of ordinal objects.

One-to-many coordination

The final feature of cardinal objects is that we can coordinate them in 'one-to-many coordinations'. Instead of coordinating one object to only one other object, as in ordinal coordinations, cardinal objects can coordinate one object to many others simultaneously. Narratively put, this occurs when we treat the different parts of something such as a sheep as all 'one sheep'.

A cardinal object does not count its ordinal parts as 1, 2 or 3, but counts each as a generic part of *one* whole. Thus it makes no sense to say that a cardinal number is the arithmetical sum of its ordinals. This definition uses cardinal counting to define cardinality itself. As Bertrand Russell rightly observes, 'counting cannot define numbers, because numbers are used in counting'.[5] Therefore, cardinal numerals, for instance, cannot be derived from the sum of their elements. This is because cardinal objects are not representations of the objects they count. They bear no resemblance to them. Cardinal objects are *different objects* than the ones they count and only coordinate with the series. In my view, we must remember that cardinal objects, like all objects, are material processes and not ideal entities.

This is also the intriguing origin of the paradox of inclusion. The paradox of all cardinal objects is that they are somehow identical with *all* of their ordinal elements, but with *none* of them in particular. The Italian mathematician Cesare Burali-Forti was the first to deduce this logical paradox in 1897. He said that if there were a set of all ordinal numbers, it would be an ordinal number greater than any ordinal in the set and thus would not be a set of all ordinals, since it did not include itself. Bertrand Russell and the German mathematician Kurt Gödel deduced a similar paradox of self-including sets. Since no set of all sets can contain itself within that set, Gödel's theorem proves there is no absolute count of elements that is not already an element different from that which it counts.

There is also a material history of this paradox in addition to these purely *logical* proofs against self-including cardinal objects. Historically, abstract cardinal objects first emerged as distinct from the elements that they 'unified'. The only real connection of a cardinal object to a counted object was the ongoing practical act of coordination. This practice does not come ready-made but required a vast material and social field of objects that I will describe in the next chapters.

But this so-called cardinal 'whole' is just another singular object kinetically coordinated to a series of other objects, 'as if' they were one whole object.[6] This same process also occurs in the invention of written language, as I have shown in *Being and Motion*.[7] The practice of coordinating a single letter to *many different* objects and sounds follows this same cardinal structure. It is no coincidence that written language and numerals both emerged simultaneously in Mesopotamia around 3500 BCE from the same material techniques. Numerals came first, and then people derived written language from the same kinetic process.

The structure of this one-to-many coordination also gives rise to the paradox of the 'last' cardinal object. A series of ordinal objects always precedes every coordinated cardinal object. The cardinal object acts as the 'final' object in the series, after which we cannot add anything more. The cardinal object then counts all the prior objects in the series as 'one'. However, as we saw above, the cardinal object that does the counting cannot be included in the 'one' that it counts. Therefore, it is not technically the 'last' object in the series because it is not in the series.

As such, there remains a fundamental undecidability as to what the last object in the series truly is. Is it the last ordinal object counted or the cardinal object that counted it? In this way, the cardinal object is uncounted and unordered in the series. Oddly then, the series is closed and unified by a cardinal object that is not in the series. What then is the true identity of the cardinal object that unifies *this particular* series? It is undecidable. The last or final cardinal object is ultimately arbitrary. Since we cannot include the cardinal in the series, the series is unaffected by whatever cardinal we use to count it.

The paradox of the 'first' cardinal is that in counting the 'parts' of a series, we are assuming already that they are parts of a whole. Parts without a whole are, by definition, not parts. For the whole to organise its parts, there must already be parts. But for there to be parts, there must already be a whole that precedes them. This produces the strange retroactive feeling that cardinal objects must have always existed.

Strangely, the same cardinal object is the *first*, *last* and *missing* object in the same sequence simultaneously. The idea of zero, which also emerged in the ancient world, is based on this phenomenon. We will discuss this further shortly.

The key idea is that cardinal objects work *as though* they precede and exceed the parts they unify. A cardinal object, like all objects, has a particular cycle. Cardinal coordination uses a specific cycle to measure many others simultaneously. Imagine a clock pendulum swinging back and forth. One-to-many coordination is like counting anything that moves in the room during this oscillation as 'one'.

This is the material and kinetic basis of coordinating cardinal objects. Anything that moves during the time it takes for our hand and brain to write the numeral 1, we can count as 'one'. Any single cycle can coordinate numerous other cycles at the same time. It is one of the most ingenious and strange moves in the history of human knowledge. Nevertheless, we should remember that cardinal coordination always requires the motion of a real material cardinal object, such as a clock, hand, word or brain. This is because cardinality is not ideal but physical and kinetic.

Sequence and simultaneity

When we use a cardinal object to count a sequence of elements, it also transforms them into equal and proportional parts of a simultaneous whole. By contrast, ordinal arrangements such as 'first, second, third' are not necessarily proportional. We might coordinate one tally mark or kinetic cycle to a bear, another to a wolf, and another to the time it takes to walk around a lake once. Each ordinal cycle is singular.

However, the same is not true of cardinal objects. A cardinal object can count any amount or kind of object as a single unit. The sequence of how we count the parts does not matter. Each becomes equally interchangeable with the others since they are all parts of the same whole.

For example, if we have a basket with six fruit pieces, we can organise the fruit into one series of six pieces, two sequences with three parts, or three series with two pieces. No matter how we order them, the cardinal number remains 'one basket of fruit'. No matter how we arrange the fruit, pears first, apples first, or bananas first, we still have 'one basket of fruit'.

After, and only *after*, we have counted the basket as 'one', we can *then* divide it into different pieces. This act of division is what creates a *cardinal series* of objects. A cardinal series forms when we add one cardinal to another. For example, we obtain the cardinal number 2 by adding 'one more' to 1. As long as this 'one more' remains identical to the first 1 in the series, the series will be proportional.[8]

Every cardinal object in the series is modelled proportionally on the first, like a copy of an original. Zero is the 'placeholder' or 'base unit' of generic wholeness. Unlike ordinal series, which change relative to each newly added object, cardinal series do not change when we add 'one more' because the model unit is self-identical. The model will only create more copies of itself.

Two cardinals are identical to one another if they share the same number of elements, regardless of what those elements are. For example, no matter how we write or think about the number 5, as long as it has five *proportional* elements, it is equal to all other sets with the same number of elements.

In this way, the cardinal series is unlike the centripetal spiral of the ordinal series, which changes as it gathers. Cardinal series are more like expanding concentric circles that move centrifugally outwards, all in proportion to the first central point. The ordinal spiral never reaches an endpoint in the centre. However, if we act as if it does, the centre can function as a model that we can use to reorganise and categorise all the objects that created it. Cardinal series are possible only if we grant the centre's identity and each concentric copy's proportionality.[9] There is nothing universal or formal about cardinal objects. Particular humans invented the structure I have described at a certain point in history through a whole field of material and kinetic process.[10] These processes are the subject of the next two chapters.

Conclusion

The incredible power and abstraction of cardinal objects makes it challenging to give a material and kinetic explanation. Nonetheless, this chapter has aimed to describe several common kinetic features of cardinal objects to prepare the reader to identify them more clearly as they occur historically in the ancient world and across several seemingly heterogeneous sciences.

In the next two chapters, I show how cardinal objects' kinetic structure emerged through the invention of numerals, accounting, measurement and logic in the ancient world.

Notes

1. Georg Cantor, *Contributions to the Founding of the Theory of Transfinite Numbers* (New York: Dover [1915]), 85: 'By an "aggregate" (*Menge*) we are to understand any collection into a whole (*Zusammenfassung zu einem Ganzen*) M of definite and separate objects m of our intuition or our thought. These objects are called the "elements" of M. In signs we express this thus: M = {m}.'
2. For a full-length treatment of this issue, see Simon Duffy, *Deleuze and the History of Mathematics: In Defence of the 'New'* (London: Bloomsbury, 2014), 141–4.
3. As quoted in Calvin T. Long, *Elementary Introduction to Number Theory*, 2nd edn (Lexington, VA: D. C. Heath, 1972), 1.
4. Cantor, *Contributions to the Founding of the Theory of Transfinite Numbers*, 85.
5. Bertrand Russell, *Principles of Mathematics* (1920) (London: Routledge, 2015).
6. For more on this, see Thomas Nail, *Marx in Motion: A New Materialist Marxism* (Oxford: Oxford University Press, 2020).
7. For the sake of eliminating redundancy the following chapters will not deal with the kinetic structure of written language, because this has already been addressed in Thomas Nail, *Being and Motion* (Oxford: Oxford University Press, 2018).
8. This is the kinetic meaning of the so-called 'universal inclusion of the void' in set theory.
9. In other words, 'the formal condition of the accumulation of objects is its opposite: the equality between all things'. Tristan Garcia, *Form and Object: A Treatise on Things*, trans. Mark A. Ohm and Jon Cogburn (Edinburgh: Edinburgh University Press, 2014), 96.
10. By failing to consider the material kinetic patterns of objects, the formal theory of objects remains ahistorical and even incorrect in its understanding of seriality and accumulation.

7

The Ancient Object I

Cardinal objects abound in the ancient world. From around 5000 BCE to 500 CE, they also became increasingly abstract. This chapter and the next argue that several ancient sciences created these cardinal objects using centrifugal motions. Specifically, I would like to show in these two chapters how cardinal objects developed out of four ancient scientific practices: the use of numerals, accounting, measurement and logic. I want to show how each of these four practices moves through four similar stages of increasing abstraction.

What I find especially interesting here is the way that each of these scientific practices moves through these shared stages and motions in its own way. There was not a linear sequence in which one method derived its cardinal objects from another. The practices emerge, then disappear, then reappear somewhere else differently. The emergence of increasingly abstract cardinal objects seemed to work more by resonance than by progressive causality and direct influence. It's more like when you strike a key on a piano in a piano shop, and all the pianos begin to resonate, than it is like knocking over a series of dominoes.

It is also important to note that the rising dominance of cardinal objects in the ancient world did not mean that ordinal objects and their patterns disappeared. Instead, they became absorbed and transformed. The accumulation of ordered objects continued, but people increasingly coordinated them to a new centralised unit to measure and divide things proportionally.

In this chapter, I focus on two scientific practices critical to the historical rise of cardinal objects: the invention of numerals in ancient Mesopotamia and the use of accounting practices in Mesopotamia and ancient Egypt. I begin first with the development of numerals.

The invention of numerals

One of the most important and earliest expressions of cardinality was the written numeral. Before numerals, we have no evidence that anyone used a single abstract unit to measure or unify many different kinds of objects.[1] In this chapter, I want to show the material-kinetic conditions of this bizarre object that remains at the foundations of scientific knowledge. Abstract cardinal numerals did not emerge spontaneously. They emerged from distinct patterns of concrete token-counting throughout thousands of years. Let's look at each of these specific patterns.

Plain tokens

What was a token? A token was a small clay object that Neolithic farmers started using around 8500 BCE to 'stand-in' for another singular object such as a specific sheep, or some particular grain, or a jar of oil.[2]

These 'plain tokens' functioned just like tallies did: by one-to-one coordination.[3] People paired one token with one singular object, not one 'kind' of object. The French-American archaeologist Denise Schmandt-Besserat has recently conducted the most extensive and pathbreaking research on token use in the emergence of numerals and letters. She writes that these plain tokens 'lacked a capacity for dissociating the numbers from the items counted: one sphere stood for "one bushel of grain", and three spheres stood for "one bushel of grain, one bushel of grain, one bushel of grain"'.[4] Each token was singular and tied to one distinct animal or object. Once that specific animal or object was dead or gone, the user discarded the token.[5] Each individual or individual Neolithic family used these plain tokens to tally up what they had centripetally gathered. People did not use these tokens for lending, reckoning, debit or credit.[6]

By 6000 BCE Neolithic farmers had begun accumulating more and more of these loose tokens. To hold them all, they started putting them in bowls, bags and jars. All these activities follow a strongly centripetal pattern of motion. As farmers accumulated more plants and animals through agriculture, their tokens accrued as well. The more they gathered, the more critical the need for a centralised storage site and container system. This centripetal accumulation was a necessary, but not yet sufficient, kinetic condition for cardinal objects.

Complex tokens and division

The first genuinely cardinal objects emerged around 4400 BCE alongside the rise of early ancient states.[7] This was the first time that ancient people started making tally marks on their tokens to count more than one other object at a time. Schmandt-Besserat calls these new kinds of tokens 'complex tokens'.

In contrast to plain tokens, people sculpted complex tokens into natural forms. Tokens that counted sheep looked like sheep heads. People then recorded tally marks, parallel lines, crosses and depressions on these complex tokens to mark the quantity of sheep.[8] The shapes of complex tokens were also more diverse than plain tokens. Instead of little spheres and cones, people shaped complex tokens into ovoids, rhomboids, bent coils, parabolas, quadrilaterals and miniature figurations of tools and animals.

However, the most novel aspect of complex tokens is that they counted more than one other object at a time. They used disc-shaped tokens to count ten animals, and cylinders to count only one animal.[9]

What is going on here, kinetically? At first, people centripetally gathered plain tokens into loose piles. Eventually, they selected one unique token to be the recording surface for tally marks. This reduced the loose stacks of disposable tokens, but it also introduced a *difference* or *division* between the single central token and the plural tally marks gathered on it. This was the first stage of 'concrete cardinality', where people coordinated one unique object with many collected objects of the 'same kind'. Complex tokens were concrete cardinals because the unit of their measure was still a specific *quality*, such as 'sheep'. For example, a large tetrahedron could not have counted or been coordinated to oil or animals, because people had already coordinated it strictly to labour.

Another kinetic aspect of complex tokens was their centrifugal movement within early cities. Temples at the centre of ancient cities were the first to accumulate the tokens since many people made offerings there. This meant temple priests were among the first to start creating and distributing complex tokens. The physical movement of tokens through the city moved from the centre to the periphery. It highlighted a social division between a few priests at the centre and many peasants at the periphery.[10]

The complex token was not merely a figurative token with some marks on it. It introduced a difference between a single token created by priests at the central temple and the plural plain tokens still used by peasants at the periphery. In this way, concrete cardinal objects were not just ideas. They were simultaneous with the movement of real physical tokens.[11]

Spheres and hierarchy

The second move towards creating abstract cardinals was when people started sealing up groups of plain and complex tokens into clay spheres, or *bulla*. According to Schmandt-Besserat's findings, these spherical 'envelopes' or 'pods' 'began being used in the Middle Uruk period, c. 3,700–3,500 B.C.'[12] and persisted in some form or another until c. 1200 BCE.[13]

What was new about these spheres was that they counted plain and complex tokens simultaneously. This entailed an even more general kind of counting. However, the major drawback of this storage system, as one would guess, was that the spherical envelopes *concealed the tokens inside*. How could one know what the sphere was holding if it hid the contents? One could verify the contents only by breaking the envelope.

So accountants began making marks on the outside of the sphere to indicate the contents. This created three levels of cardinal counting. At the first level, the complex token counts the ordinal tally marks on its surface. At the second level, the clay sphere counts all its tokens as 'one' sphere of tokens. Then, at a third level, the marks on the sphere count the individual tokens inside.

However, these three counting levels were all still *concrete* cardinals, since 'the markings only repeated the information encoded in tokens for the convenience of accountants'.[14] A concrete cardinal of a concrete cardinal is still concrete because people tied it to *a specific quality*. For example, if there were six tokens in the sphere, the same token would be impressed six times on the outside.[15]

There are two critical kinetic consequences of using these spheres. The first is that they demonstrate clearly that cardinal objects assume a prior accumulation of ordinal objects. For example, before central priests could officially distribute tokens, they had to be accumulated into bowls, jars and bags by peasants. Once people gathered the tokens in these open vessels, the vessels were sealed to create an enclosed unity: the sphere. Cardinal unity was not an idea first. It was a real kinetic and material act of sealing up tokens in a physical clay sphere.

It was also significant that people made the clay spheres by a centrifugal method. First, one formed some clay into a solid ball, then poked a hole in it and started working a hollow into the middle *from the inside out*.[16] Finally, a plug sealed the contents and covered the tracks of the process that shaped it. Thus, the open vessel preceded the closed sphere, just as centripetal gathering preceded centrifugal radiation.

The second kinetic consequence of the token sphere was to introduce a kinetic *hierarchy* between the invisible centre of the sphere and the visible periphery. For example, by sealing up the clay vessels with an official cylinder-seal, the elite priest-bureaucracy granted superiority to the tokens inside as unseen models copied by the peripheral and imperfect impressions on its surface. People trusted the seals' authority to count the hidden tokens inside because priests made the seals with authorisation from the gods. The unseen tokens' superiority was very much akin to the unseen gods' supremacy over the visible world.

Tablets and reorganisation

The third move towards creating abstract cardinal numerals occurred when people stopped putting tokens inside the spheres and instead just impressed them into the soft clay. Once a system of notation was invented for impressing the outside of the hollow spheres, storing the tokens became superfluous. Instead of using hollow spheres, people started making impressions on solid spheres and flattened spheres that were more like convex tablets. This transition happened over about two hundred years from c. 3700 to 3500 BCE.[17]

However, between c. 3500 and 3100 BCE, people still impressed tokens in one-to-one coordination on the surface of these solid spheres and tablets. This meant that the tokens' unique shapes were still bound to particular kinds or qualities of the things people counted, such as sheep, oil or grain. In other words, concrete cardinality still prevailed.

Around 3100 BCE Mesopotamians made another striking innovation to the counting process. Instead of impressing tokens into the clay tablets' wet surface, they started using sharp sticks to incise or draw the token shape on the clay. That way, instead of having to make and use a token for each impression, they could draw an image of the token or a pictographic sign coordinated to that token.

This pictographic shift led to a further transformation of the cardinal object. Instead of either impressing or drawing nine '1 sheep' tokens, they started to designate a particular token as the 'quantity token' to be impressed or drawn alongside the sheep token. This saved room on the clay surface and conserved labour on how many tokens scribes had to draw. For example, around 3100 BCE people began to use a grain token impression next to an incised pictograph of a sheep to indicate the quantity of sheep, *not the amount of grain!* This was an incredible event. For the first time, a unique kind of token, the grain token, was treated as a pure quantity as if it had no quality. Inversely, people treated the sheep pictograph as a pure quality without any quantity.

This revolutionary beginning of the first numerals was as strange as it was powerful. It was the first time people started *acting as if* quality was separate from quantity – something found nowhere in nature. In this way, the object's quality and quantity were split up and reorganised into different areas on the tablet's surface.

Kinetically speaking, what is essential to see here is that splitting up and reorganising quality and quantity was only possible once people started acting *as if* the tokens were at the centre of the sphere. Once people were dealing with imaginary tokens, it felt possible to imagine them with just their quantitative or qualitative aspect. In this way, the imaginary centre of the sphere allowed for a *relative abstraction* and *reorganisation* of quality and quantity. However, people were not yet dealing with absolutely abstract cardinal objects because they were still using *grain* tokens and pictographic incisions that were visually tied to certain qualities.[18]

The final transition to using numerals was by far the most radical and abstract.

Script and cardinal abstraction

The fourth and final move in creating cardinal numerals occurred when ancient Sumerians stopped using grain tokens *as if* they were pure quantities and began using incised numerals. Since, in the third step, the tokens were only *imagined* to be inside the sphere, then any token could be used to quantify any quality, such as using grain tokens to quantify sheep tokens. In the fourth step, however, this conclusion is radicalised. If people only imagined the tokens, then *any* mark can quantify any token whatsoever.

This is the most radical and abstract conclusion of the whole kinetic process. Any object can be used as a cardinal object to count any other object. This kind of abstraction was only possible because material token objects can be moved and concealed inside one another. Abstract cardinals may seem like immaterial ideas, but they are ultimately habits of acting based on ordering objects in specific centrifugal patterns.

Around 3000 BCE the ancient Sumerians starting using a wedge-shaped stick to make indentions called 'numerals' in wet clay and other materials. Unlike tallies, tokens, spheres or tablets, people treated numerals as pure quantities from the beginning. This new cuneiform script did not immediately replace pictographs and token use but existed alongside it. By 2900 BCE the Sumerians had begun using cuneiform writing to record abstract quantities without qualities (numerals) and abstract qualities without quantities (written language).[19]

The invention of numerals also introduced a whole new way of counting a series. The cardinal numeral 2, for example, is not just 'another numeral after' the first one but is *precisely* one more than 1. Two is not only *after* 1 but is a new cardinal object that includes 1 and all the coordinated elements of 1. Every cardinal number is thereby *precisely proportional* to the first cardinal number.

The first abstract numeral 1 was the first and central model of a unity upon which all other numerals were proportionally based. In this way, cardinal counting is a centrifugal act that begins with the imaginary pure quantity at the centre of the clay sphere and sees each following numeral as a proportional instance or incarnation of that unseen object of pure quantity.

The abstract cardinal numeral became the centre and model by which people could centrifugally count all other peripheral quantities.

Once this kinetic process emerged in ancient Sumer, it spread around the world up to the present. Although people tend to treat numerals as either arbitrary or eternal, I hope I have shown that their genesis is entirely material and kinetic. Each time we use cardinal numbers, we repeat this ancient habit grounded in a centrifugal motion pattern.

Accounting

A second significant contribution to the development of cardinal objects in the ancient world was the invention of accounting. What is accounting? Accounting is different than counting. Counting describes or *records the order of objects*, but accounting *reckons and sums them up*.[20] Accounting is a meta-count of something that someone has already counted. However, it is also a recount that counts in a *different way* than the first count.

Accounting does not just gather objects centripetally into an ordered series. It uses particular central or privileged objects as units of measurement. Despite the heterogeneity of a group of objects, one can account for them using the same meta-count or *account*. Accounting involves not just a 'final' object in the series that 'sums' the whole series, but includes a first object that is the unit of measure for counting each object. In this way, accounting can coordinate any series of objects whatsoever to a single object. Accounting is also free to change the unit of how things are counted or measured.

The archaeologist Schmandt-Besserat distinguishes between two kinds of counting: computing and accounting.

> Computing consists of making calculations. Accounting, on the other hand, entails keeping track of entries and withdrawals of commodities. It is my contention that, computing (of time) took place within egalitarian societies, but that the origin of accounting must be assigned to ranked societies and the state. In other words, it was not simply hoarding grain or producing manufactured goods that brought about accounting: social structures played a significant role.[21]

The rise of numerals, money, debt and taxation are interrelated, and 'all emerged in the context of controlling goods, stocks and transactions in the temple economy of Mesopotamia'.[22] Once ancient Sumerians invented accounting in Mesopotamia, it spread through the ancient world, including Egypt, Greece and Rome.

In this section, I want to show how several primary accounting techniques of the ancient world came about following the same four steps of cardinal abstraction described in the invention of numerals. In particular, I want to show how the creation of money, debt, taxation and redemption all relied on the centrifugal movements characteristic of cardinal objects.

Money and division

One of the earliest tools of accounting was money. What is money? Money is not a single object or even a particular type of object. Money is a performative relationship between objects. Money is how objects circulate between people. Historically, money emerged in Mesopotamia around the same time as the practice of using cardinal numerals. Money was a cardinal unit of measure chosen by the centralised priest-bureaucracy of Mesopotamia to secure a system of equivalencies between very different objects. The first selected central object of securing this system was a specific weight of a piece of silver called a 'shekel'. In this system, one silver shekel equalled one *sila* of oil = one *sila* of grain = one *mina* of wool = one *ban* of fish.[23]

However, money is not a static object or even a weight. People have used many different objects throughout history as currency. What remains common to all these systems of money is the system of general equivalence itself. Accounting records this system.

Money is a system of changing exchange ratios between cardinal objects. Instead of using a single cardinal object to coordinate the motions of any series of objects, as numerals do, money establishes an equivalence of exchange between several different *cardinal* objects *regardless of their numeric quantity*.

For example, one silver shekel could be equal to 1 or 2 *sila* of oil, depending on the harvest. The numeral 1, by contrast, is only ever equal to 1. Objects change, and equivalencies change, so accounting is the act of creating and adjusting equivalencies between different objects. Crucially, accounting sets the rates of exchange by standardising a system of weights and measures. For example, the fixed weight of a silver shekel determines its value.

Kinetically, accounting's first kinetic operation was this *division* between a *centralised* system of weights and measures and the periphery of different objects that people exchanged. The priest-bureaucrats in the central temples of the city were the first accountants. The more their activity was standardised and centralised in the city and the temple, the more objects from farther away people could exchange in the city. Having a centralised exchange and accounting system made it possible to unite and administer trade among different peoples across different geographies. Accounting standards held together empires.

However, the centralisation of accounting standards was only possible because diverse objects at the periphery were mobile enough to be transported to the city. We must remember that peripheral people and movements continually sustained the centre and its centrifugal administration. In this way, there was a feedback loop or circulation between the centripetal accumulation of objects and the centrifugal application of a standard of exchange rates between objects. Portable tokens, spheres and tablets helped make accounting a mobile activity as well.

For example, the wider the empire spread, the more massive temples and city-states became, and the more their periphery grew disproportionate to the size of their centre. This required faster, more mobile and longer chains of coordinated objects. The more objects people can coordinate from the periphery to a single system of master cardinal objects such as linguistic scripts, numerals and economic equivalence, the more powerful the empire.[24]

Interest

Another early aspect of accounting was interest, which was a portion of the money paid on loans and debts. In ancient Sumer, only certain city people such as temple priests and private urban merchants charged interest.

These people used interest to adjust the existing money system of cardinal objects and their equivalence. Because of this, credit and debt worked very differently in ancient societies than in prehistoric ones. In most Neolithic cultures, ordinal objects were never equivalent, and so exchange, credit and debt were always unequal and disproportionate. People 'owed' one another 'something else', but there was no accounting or equivalence between what someone gave and what someone else gave them back. It was like buying lunch for a friend who will buy you lunch next time and so on back and forth, without anyone considering the price of the meals.

As more and more people gathered centripetally in cities or rulers forcibly enslaved them, urban populations grew. Palaces emerged in the centre near the temples and demanded donations, offerings and taxes from their peripheral territories. The priests of ancient Mesopotamia began to use this surplus to provide the earliest recorded loans. By 2400 BCE the initial *division* between central standards of exchange and the exchanged peripheral objects had become a clear *hierarchy* of lenders over debtors. Lenders could request an extra amount of money above the standard cardinal exchange rates, and debtors by law had to pay it. Money established an equivalence between two *different* regions, the centre and the periphery, but interest established a *hierarchy* between them.

Here is the fascinating aspect of interest that I would like to point out. Interest was not an additional amount of money added to the regular exchange rate. Instead, it was a change in how the objects exchanged were counted. One of the most perplexing aspects of cardinal objects is that they can always count things differently since their unities are arbitrary.[25]

For example, let's say a Sumerian temple priest lends a box of silver shekels to a merchant, who will travel abroad and exchange with them. Let's call this box of silver shekels 'cardinal object A' and the series of individual silver shekels in the box 'ordinal elements a, b, and c'. When the merchant returns to the city, he will owe the priest-accountant a debt of one box of silver shekels. Let's call that 'box B'. Qualitatively, the two boxes will not be the same, but because of the cardinal system of equivalencies, or 'money', the priest and the merchant can agree to pretend they are: $|A| = |B|$.

However, because one can count the elements of cardinal objects differently, the priest decides to recount the contents of box B by pairs so that each shekel is *one* 'pair of shekels'. We can write this as:

$$A \{a, b, c\} \rightarrow B \{2a, 2b, 2c\}$$
$$|A| = |B|$$

Since cardinal equivalence is based entirely on the bijective (one-to-one) coordination of its elements, i.e. $\{a{\rightarrow}2a, b{\rightarrow}2b, c{\rightarrow}2c\}$, these two cardinal boxes *are equivalent*, even though they are monetarily *double*, and hence unequal for the debtor. This seems odd, but the counting preference of the hierarchically superior priest was agreed to by the travelling merchant. It is no more bizarre or arbitrary than deciding that box A and box B, both filled with qualitatively unique shekels, are the same. Nor is it any more odd than saying that box A = box A + 5 shekels. It is strikingly apparent that box A cannot be quantitatively equivalent to another box with 5 more shekels. However, if one counts the shekels in pairs, we can say box A and box B can be coordinated one-to-one and thus *counted as* equivalent.

Since cardinality is simply coordinating one object to many others, and not a timeless, immutable essence, cardinal coordination can change over time. This was the basis of interest.

Taxation and reorganisation

A third significant aspect of accounting was the invention of taxation. Similar to interest, taxation was a change in the way that people counted objects. However, instead of being limited to interest-bearing loans, taxation first emerged when accounting changes were applied centrifugally to the *entire* periphery by a single centre. In this way, taxation relied on the state's central authority to temporarily reorganise how all objects were counted and exchanged.

Contrary to primordial debt theories, taxation was not an absolute debt to a god or king. Taxation did not *derive* from religion or kings, but rather religion, kings and accounting all followed and co-emerged through the same centrifugal patterns of motion.[26]

Just as priests did not add interest to monetary exchange, accountants did not add or subtract taxes. State accountants merely recounted the object of taxation as more or less than it was before taxation. Taxation changed the relations of equivalence between objects. In this way, the process of taxation is not significantly different than money itself. Both use cardinal counting, but in taxation, the state temporarily transforms general equivalence standards for all objects.

Whoever controls the accounting controls the relative equivalence of cardinal objects. Therefore, money, interest and taxation do not add or subtract but instead adjust the equivalence relations differently. These three accounting techniques all use the mobility and arbitrary nature of cardinal counting in different ways.

Taxation was rarely direct in the ancient world. In Mesopotamia, Egypt, Greece and Rome, for example, taxation took the form of a periodic fluctuation or fee on specific objects and services such as property, cattle and grain. Taxation was not a general or absolute debt to god or king. Therefore, taxation did not require coinage, since money is just a standard of equivalencies in accounting. In Mesopotamia and Egypt, accountants enforced taxation when they registered transactions. Taxation effectively meant that the taxed cardinal object was merely accounted differently than it was before.

Periodically, the Egyptian king and his court of scribes and accountants would set out across the kingdom, accounting everything using various standardised weights and measures. They accounted property, grain, cattle and other objects. Critically, taxation coincided with accounting. Since the grain grown by the farmer had no absolute measure independent of the royal account, there was not technically any subtraction of a tax.

If this sounds like an enormous scam, it is. There is a real difference between more or less grain for the people who need it to survive. However, cardinal accounting's power is to treat the grain as a pure quantity without qualities or use-value. No matter how large or of what type, an accountant can easily count any set of ordinal elements as 'one' cardinal sum. It just depends on what the cardinal unit of measure is and who is accounting. In the case of taxation, it is the centrifugal movement of the state that accounts its periphery.

Redemption and abstraction

The fourth and final step in the cardinal accounting process was the most dramatic and abstract of all: the practice of redemption. In contrast to the above three techniques of concrete cardinal accounting, redemption abolished all changing equivalence systems based on various political interests in favour of a single and unchanging cardinal equivalence. Redemption was the most abstract because it introduced a new form of debt that no one could concretely repay.

The effect of concrete accounting techniques such as money, interest and taxation was that ancient administrative centres were continually extracting a larger and larger surplus from their peripheral populations. Eventually, when enough of these populations ended up in debt slavery, it undermined agricultural production, depopulated the countryside and created revolts that negatively affected the state. This is why every ancient empire periodically engaged in some form of jubilee or debt forgiveness. Forgiveness was a way of resetting the accounting process. This redemption act was something that no one could concretely repay, which is precisely the point. Redemption occurred when accounting reached the practical limits of its theft from the periphery. The debt slave then owed his or her freedom and life to the state, something that could only be paid back in death. Typically, this meant death in battle as a military conscript.

However, the most radical and abstract form of redemption eventually took place when the Semitic peoples imagined that a single divine king had permanently abolished the whole process of periodic redemptions and earthly systems of equivalence. Instead, the freed slaves of the accounting cycles would owe an impossible debt to an invisible cardinal object: God or Yahweh. The Hebrews imagined Yahweh as a divine accountant whose count was absolute and total, similar to the invention of the numeral 1. God counted all his chosen people, the Israelites, as one and thus freed them from all arbitrary earthly systems of servitude and equivalence. By redeeming the Israelites, Yahweh abolished all their debt, liberated them from Egyptian slavery, and gave them land.

In one sense, Yahweh temporarily abolished all earthly accounting and debt. Still, in another sense, he introduced an even more profound debt owed to a single consolidated and abstract cardinal object: himself. According to the Hebrew Bible, Yahweh created the world and promised to redeem God's people eventually. However, in the meantime, he became an absolute landlord and abstract king to whom the Israelites owed complete obedience. No concrete sacrifice on earth would ever be enough to pay back the creation of life itself.[27]

In this way, I hope that it is clear that there is no such thing as 'debt in general'. Debt functions differently in different fields of knowledge. Ordinal debt, for example, means that someone owes *something* more. However, concrete cardinal debt means that someone owes an exact numerical quantity of something, but that the structure of equivalence can change if accounted differently. Ultimately, the idea of an abstract cardinal debt that no one can repay reveals the radical non-equivalence that lies at the heart of cardinal equivalence itself. The kinetic act of cardinal counting itself is arbitrary and not equivalent to any specific cardinal object.

Conclusion

This chapter aimed to show that four stages of centrifugal motion in the ancient world shaped cardinal objects' historical emergence. In particular, I showed how the invention of numerals and accounting techniques helped create these cardinal objects. In the next chapter, I want to continue this line of argument and show how two other major ancient sciences also contributed to the emergence of cardinal objects: measurement and the invention of logic.

Notes

1. Georges Ifrah, *The Universal History of Numbers: The Computer and the Information Revolution* (New York: Penguin, 2005).
2. Denise Schmandt-Besserat, *Before Writing* (Austin: University of Texas Press, 1992), vol. I, 6.
3. Beginning around 8000 BCE and continuing through five millennia, the tokens were hand-moulded out of clay, and sometimes afterwards baked in an oven or fired in a kiln. In later periods a relatively small number were cut out of stone. Schmandt-Besserat, *Before Writing*, vol. I, 198, 20–31.
4. Denise Schmandt-Besserat, 'The Earliest Precursor of Writing', in David Crowley and Paul Heyer (eds), *Communication in History: Technology, Culture, Society*, 5th edn (Boston: Pearson, 2007), 14–23 (19).
5. Tokens were singular, unique and destroyed after use. Denise Schmandt-Besserat, *How Writing Came About* (Austin: University of Texas Press, 2006), 93. 'In other words, they were used primarily for record keeping rather than for reckoning.' Schmandt-Besserat, *How Writing Came About*, 30.
6. Schmandt-Besserat, *How Writing Came About*, 30.
7. Schmandt-Besserat, *Before Writing*, vol. I, 24–5, 198.
8. Schmandt-Besserat, *Before Writing*, vol. I, 14, 82.
9. Schmandt-Besserat, *How Writing Came About*, 115.
10. Schmandt-Besserat, *Before Writing*, vol. I, 91. In general, (complex) tokens were associated with public rather than private buildings.

11. See Schmandt-Besserat, *How Writing Came About*, for images of complex tokens, frequently with holes or numerical markings in the centre.

12. Schmandt-Besserat, *How Writing Came About*, 44.

13. Schmandt-Besserat, *Before Writing*, vol. I, 198.

14. Schmandt-Besserat, *Before Writing*, vol. I, 154.

15. Schmandt-Besserat, *Before Writing*, vol. I, 198–9.

16. Schmandt-Besserat, *Before Writing*, vol. I, 112.

17. Schmandt-Besserat, *Before Writing*, vol. I, 133.

18. Schmandt-Besserat, 'The Earliest Precursor of Writing', 21.

19. Schmandt-Besserat, *How Writing Came About*, 21. For a more in-depth account of the kinetics of writing with respect to ontology, see Thomas Nail, *Being and Motion* (Oxford: Oxford University Press, 2018). With respect to literature, poetry and drama, see Thomas Nail, *Theory of the Image* (Oxford: Oxford University Press, 2019). Here we deal only with the quantitate dimensions of cardinality.

20. From Middle English, from Anglo-Norman *acunte* ('account'), from Old French *aconte*, from *aconter* ('to reckon'), from Latin *computō* ('to sum up').

21. Schmandt-Besserat, *How Writing Came About*, 103.

22. Keith Robson, 'Accounting Numbers as "Inscription": Action at a Distance and the Development of Accounting', *Accounting, Organizations and Society*, 17.7 (1992), 685–708 (699).

23. Samuel Kramer, *History Begins at Sumer* (Philadelphia: University of Pennsylvania Press, 1988), 260.

24. See Robson, 'Accounting Numbers As "Inscription"'.

25. See Karl Marx, *Capital Vol 1* (New York: Penguin 1976), 247, on the difference between Commodity–Money–Commodity and Money–Commodity–Money.

26. See Thomas Nail, *The Figure of the Migrant* (Stanford: Stanford University Press, 2015), and Nail, *Being and Motion*.

27. 'You shall do this in the following manner: the creditor shall pardon any debt of his neighbour or brother, and shall stop exacting it of him because Yahweh's pardon has been proclaimed' (Deut. 15:2); 'the land that Yahweh gives you' (Deut. 18:9); 'Remember that you too were a slave in the land of Egypt, and Yahweh, your God, has given you freedom. Because of this, I give you this commandment' (Deut. 15:15); 'You shall consecrate to Yahweh all the male firstlings that are born of your cattle or sheep. You shall not use the firstling of your cattle for work, nor shear the firstling of your sheep' (Deut. 15:19); 'Even if someone is not redeemed in any of these ways, they and their children are to be released in the Year of Jubilee, for the Israelites belong to me as servants. They are my servants, whom I brought out of Egypt. I am the Lord your God' (Lev. 25:55).

8

The Ancient Object II

In this chapter, I want to show how two more significant scientific practices contributed to developing cardinal objects in the ancient world. In particular, I want to show how the creation of new measurement techniques and the invention of logic followed a similar four-stage process towards increasing centrifugal movement and cardinal abstraction.

Metrology

Metrology is the study and use of measurement. Measurement is the application of a central cardinal unit of measure to a periphery of diverse objects. Beginning around 3500 BCE, ancient Sumerians invented a standardised system of weights and measures and used it for various purposes. But before we get into its uses, we need to define what a 'unit' of measurement is.

The unit of measure and division

A unit of measurement is a cardinal object that people use to compare the quantity of one thing to another, proportionally. Therefore, this kind of measure assumes a kinetic division between a unique object to be the unit and a range of other objects that one will measure with the unique object. This was a genuinely new use of objects.

For example, prehistoric measurement dealt only with non-numerical quantities. In prehistoric science, every object was relative to every other. Measurements of length, volume, weight and time were only 'more or less' than one another. There was no unit of difference between objects. A sapling was smaller than a tree, a river was larger than a creek, and a rock might be larger or smaller than a human. Archaeologists have even discovered a series of smaller and larger weight-objects from as late as the Iron Age. These were made of stone but had no discernible proportional ratio between their weights. This finding strongly suggests that prehistoric people did not use them as cardinal units of measure.[1]

The historical origins of cardinal measurement are intensely kinetic. In ancient Sumer, cardinal measurements began with the first attempts to measure grain movement into and out of the city's central granary. Once a harvest of grain was centripetally gathered into the granary, it was measured by units and centrifugally distributed back out to the periphery as needed. The historical emergence of the earliest standards of weights and measures was motivated by the mobility of grain and grain tokens.

Early Sumerian priests accomplished this by establishing a unit of measures such as the weight of a silver shekel or a *sila* of grain, administered by a central bureaucracy. The material and kinetic properties of grain and silver allowed them to become the first units of measure. Silver can be readily melted into a fluid and reformed into various solid states. Large quantities of grain move like fluids and administrators can portion them out into different-sized containers. Silver and grain are relatively homogeneous materials that can also be well preserved over time.[2] They even have practical use-values for making jewellery and eating, respectively.

It is no coincidence that people derived the first units of spatial measurement from concrete *mobile* objects: a grain, a finger, a foot, a cubit (forearm), a step, a reed, a rod and a cord. For an object to become a unit of measure, its mobility and stability are crucial. The unit cannot be immobile, or else one cannot take it elsewhere to measure anything. However, if the object has only a fluid state like water, it will move too readily and fail to measure anything. The cardinal-unit object must be portable, reproducible and relatively homogeneous. After silver and grain, the human body was the next highly mobile and proportional body that ancient peoples used as a unit of measure. After the human body, they used sticks and ropes.[3]

In ancient Egypt, measurement emerged as a response to the movements of the Nile. When the river flooded, it concentrated nutrients in certain areas. The problem was how to measure out and distribute these nutrients through farming and taxation.

Centrifugal measurement and hierarchy

Once ancient peoples had selected a particular unit of measure, they applied that unit in a centrifugal motion to other objects. The priests and scribes who picked and used these units lived in urban centres and were socially superior to the people and objects they measured. Numerals, accounting and units of measure all emerged around the same time because they were invented and applied by the same centralised priest-bureaucracy.

Kinetically, measurement began with the selection of a concrete unit-measure object from the city-temple-granary. Then, someone physically walked out to the city's periphery or into the countryside and used this object to measure others. The unit-measure was decided in the centre of the city and extended radially outwards in every direction.[4]

For example, a length was measured by first determining a unit-object, placing it at a fixed starting point, and then extending it one measurement at a time outwards from the starting point. Each time the measurer stretched the unit, they treated it as another cardinal object added to the previous one. As the measurer continued, they learned more about the length of the object they were measuring.

In ancient Sumer and Egypt, weight was determined centrifugally by a balance scale. Equality occurred when weight was distributed evenly across a central point. The balance scale dramatised the relationship between social justice, economic exchange and measurement exquisitely. All were secured and ordered around a central cardinal object: the state. For example, from around 2100 BCE onwards, there were Sumerian tablets with hymns to a goddess named Nanshe, who presided over social justice, exchange and measurement. On these tablets, the Sumerians thanked Nanshe for the standardisation of the reed basket size, the fair distribution of grain, and the law that secured the distribution.[5]

Several Mesopotamian tablets and stele depict various gods giving a measuring rod and rope to a king. At the top of the Code of Hammurabi stele, which contained all the state laws, the sun god Shamash gives the rod and cord to Hammurabi. On the Ur-Nammu stele, the moon god Nannar provides the king with a rod and coiled rope. An Old Babylonian Burney relief (2300–2000 BCE) shows the goddess Ishtar holding out the same rod and rope.

In these images, the centrifugal structure of measurement was explicit. The perfect unit-object was created in the heavens by the gods and handed down to the holy mountain centre. The king extended the measuring unit outwards and downwards in all directions towards the periphery of his people and lands. In this way, the periphery was socially and metrically dominated by the centre.

Arithmetic and reorganisation

Measurement was also the basis of arithmetic.[6] The oldest known piece of recorded mathematics is a tablet found in Uruk dating from around 3200 BCE. The writing on the tablet estimated the size of a field by multiplying the average of the lengths of the sides AB and CD by the average of the lengths of the sides AD and BC.[7]

To measure was to move a unit-object along the surface or side of another object, like unfolding a coiled rope. Arithmetic emerged from the mobile act of measuring or walking. This was explicit in the hieratic

Figure 8.1 Stele of Hammurabi, Louvre, Paris. Author's photograph.

Egyptian symbol for addition: legs walking in the direction of the equation. The sign for subtraction was legs walking in the opposite direction. In Egyptian writing, the numerical mark for 1 was a picture of a measuring rod, evoking unity and identity. The number 10 was the yoke of a plough, and 100 was a measuring rope. All of these arithmetical and numerical images were associated with the movements of measurement.

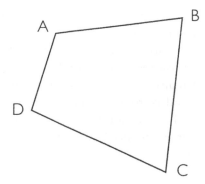

Figure 8.2 Uruk field area. From Norman Biggs, *Quite Right: The Story of Mathematics, Measurement and Money* (Oxford: Oxford University Press, 2016), 19.

Arithmetic and addition are quite different than merely counting cardinal numerals in order, 1, 2, 3. When we count cardinal numbers, we are not adding anything. We do not need to know that 1 + 2 = 3 to count to three. All we need to know is that the numeral 3 is a cardinal unity that contains three ordinal elements.[8]

However, the foundation of addition was the practice of measurement, where it was common to count by units other than 1. In arithmetic, the cardinal number line becomes the object of measure. The brilliant move of addition was to *vary the unit of the count itself.* All arithmetic is nothing more than measuring the cardinal number line by different unit-objects. For example, if we first move five steps forward, then 11 steps forward, then 16 steps forward on the number line, any measurer can tell you that you have walked 32 steps or units forward. The same sum is reached even if the units of the count vary: (5 + 11 + 16 = 32) = (3 + 14 + 15 = 32).

Furthermore, the basic properties of commutativity ($a + b = b + a$) and associativity ($a + b$) + $c = a + (b + c)$ are also effects of the metrological nature of arithmetic. Whether one walks a measuring rope forwards or backwards does not affect the measurement (commutativity). Nor does it matter if one walks partway, adds the sum, and then walks and adds further along. *Numerals are not conceptual self-identities*, as Russell and Frege wrongly thought,[9] but kinetic and metrological *movements.*

Subtraction was derived from addition.[10] One could decompose a whole by subtraction or fractions only if one composed that whole by addition in the first place. Logically, the whole and its parts are conceptually co-given. Still, historically and materially, the measure of a whole must first be made through motion and measurement before one can subtract anything from it.[11]

Multiplication emerged from the iterative act of using the same unit again and again to measure another object. The unit-object was folded, du*pli*cated or multi*pli*ed relative to another object. We can see this logic of folding in knotted measuring ropes found in Mesopotamia and Egypt. A cord can be knotted at any interval to reflect the iteration of any unit of measure. The knotted coil of rope visually shows the folded and kinetic structure of arithmetic. The measuring cord renders visible the material continuum, which underlies and holds together all arithmetical operations: duplication and unification.

Figure 8.3 Egyptian numerals and arithmetic. Author's drawing.

In the Ahmes Rhind papyrus, multiplication problems were solved by successively adding copies of the multiplicand. The multiplication of 69 by 16, for example, would be performed by adding 69 to itself to obtain 138, then adding this to itself to reach 276, applying duplication again to get 552, and once more to obtain 1104, which is, of course, 16 times 69.[12] Division was accomplished by reversing the process and simply doubling the divisor instead of the multiplicand. For example, 100 divided by 5 is performed by adding 5 to itself to obtain 10, then adding this to itself to reach 20.[13]

Arithmetic allowed the ancients to develop a plan or model of possible measurements and their relations before putting that plan into action. Before one built a temple, for example, one could make an arithmetical plan of the temple and then fill in the concrete objects afterwards. The arithmetical plan not only introduced a mathematical model but applied this model to the earth, producing a concrete copy of the model.

For example, ancient Babylonians used an arithmetical plan to design a grain-storage pit of a standard size such that it was divisible equally by *sila* grain vessels (about a litre) for distribution. Records on clay tablets suggest that the grain pits were one rod (about 6 metres) square and one cubit (about 50 centimetres) deep. To divide the *sila* evenly, they had to be 180 cubic fingers, so the vessel's height had to be 180 divided by 27, or six and two-thirds fingers.[14]

Also, before digging a canal, Babylonian engineers would plan the entire project arithmetically. They would calculate the trapezoidal cross-section, the volume of earth to be moved, how much a digging man could do in a day, and come up with the number of working days needed for the job.[15]

The Egyptians seldom made measurements before they laid out a mathematical plan. Arithmetical calculations such as the Pythagorean theorem and the area of a circle were worked out and recorded on ready-to-hand tablets before the building began. Kinetically, arithmetic made possible the periphery's reorganisation using a central set of arithmetical models and tables.

Algebra and abstraction

The fourth and most abstract cardinal object was the invention of the algebraic variable. A variable is a cardinal object whose value can be any value and whose unit can be any unit. A variable is an *unqualified* and *unquantified* quantity. By contrast, in arithmetic, one does something with two or more *known* numeral quantities or qualified measurements to create a result. The critical innovation of algebra was not just the introduction of an abstract object with an unknown quantity into arithmetical equations. Algebra allowed people to act *as if* all determinate objects were determinations of more primary generic quantities. It was as if the world was purely abstract forms that people determined through measurement, and not the other way around. People even treated abstract numerals as local manifestations of an even more abstract and undetermined quantity. It is hard to underestimate how radical this move was. It was the first use of an unqualified and unquantified object in the history of science.

In ancient Egyptian algebra, the term for this abstract cardinal variable was *aha* or 'heap'. A heap was not a unit of measure but rather an object coordinated to an unknown quantity of elements. A heap was related to the real presence of a centripetal accumulation of objects and a centrifugal desire to distribute them.

These 'heap problems' were solved by what Western Europeans would later call 'the method of single false position'.[16] One could substitute a heap for any determinate value as an approximation for the solution. If the answer was not correct, the approximate heap value could be substituted again more accurately and so on until the equation was solved correctly.

This method made explicit the mysterious nature of the variable's pure quantity and how one could substitute any concrete number or quality for any other. This, I believe, is what the Egyptian mathematician Ahmes had in mind when he wrote that 'Accurate reckoning [is] [t]he entrance into the knowledge of all existing things and all obscure secrets.'[17] 'All existing things' referred to the known numerals and their arithmetical relations, while 'all obscure secrets' referred to the abstract algebraic variable's mystical nature.

Logic

The fourth scientific practice that contributed to the development of ancient cardinal objects was the invention of logic. Although we often think of logic as a purely mental activity, it has a material basis in history and follows a similarly centrifugal pattern to other ancient sciences.

We can hear the kinetic roots of logic in its ancient names. In Old Akkadian, scribes concluded logical and mathematical problems with the phrase 'such is the procedure [*nēpešu*]'. The word *nēpešu* meant 'to lead a religious procession', 'to use a tool', 'drive a siege machine', or 'a construct a building'. In ancient Egyptian, the word used for logical calculations was *sSmt*, meaning 'to lead a procession'. For these scribes, logic was a movement that built or did something, not merely a mental operation.

In Greek, the word *logos*, from which we get the word 'logic', comes from the root word λέγω, *légō*, meaning to 'to pick up, gather together, to choose, to arrange'. Contained in Greek logic's archaic roots was the *kinetic activity* of picking up, selecting, gathering, drawing together and arranging of thrown objects. In Latin, the word *dūcō*, from which we get the logical words 'abduction', 'induction' and 'deduction', means 'to lead, guide, draw, or pull'. Ancient writers did not speak of logic as static, but as a movement or process.

Let's look at the four historical movements of logic that contributed to increasingly abstract cardinal objects.

Abduction and division

The first logical operation of the ancient cardinal sciences was abduction. Abduction used empirical observations to 'draw' (*dūcō*) conclusions between cause and effect. 'Such knowledge', according to the American logician Charles Sanders Peirce, 'must involve additions to the facts observed [and] may be called a hypothesis.'[18] Abduction happens when we see many things happening, but then we imagine another object we call 'causal force'. This causal object is like a one-way arrow between one object, the cause, and another, the effect. Kinetically, this object introduces a division between a central cause and its peripheral effects.

We often think of this kind of logic as universal, but it is not. It emerged historically alongside other cardinal objects. Before this unilateral notion of causality, prehistoric humans did not necessarily believe in hypothetical or invisible objects of causality. There was not a single causal object but rather an assortment of objects, all with agency. In an ordinal sequence, for instance, each new ordinal changes the whole. Ordinal causality is immanent and collective.

For example, prehistoric people did not necessarily see 'rain' as the sole causal object or reason why 'the grass is wet'. When it rained, the whole world changed. The sky darkened, *and then* the temperature dropped, *and then* there was moisture in the air, *and then* the grass was wet. It was a sequential logic where everything that came before played a part in what came next. We might call this ordinal logic 'transductive' because causality 'moves through' everything immanently.

Abduction, however, *added* an *external hypothesis* to explain or predict more than one series of objects. Causality worked as if it were an invisible object attached to other concrete, sensible objects. Interestingly, though, the decision to make one object the cause of another is just as arbitrary as counting many different things as 'one'. As Peirce says, 'No reason whatsoever can be given for it, as far as I can discover; and it needs no reason since it merely offers suggestions.'[19] Nature is not pre-grouped into universal units or linear causal sequences. Yet, according to Peirce, 'every single item of scientific theory which stands established today has been due to Abduction'.[20]

Historically, people first started using abduction when they wanted to control the homogeneous reproduction of events. For example, Mesopotamian cities and states began writing causal histories, instruction books, recipes and calendars to repeat events more precisely. With the emergence of these written documents, causality seems to take on a life independent of sensible objects. The written rule, ritual and recipe were hypothetical causal objects whose *procedures* ensured other objects' causal order.[21]

Abductive logic attempts to govern the order of things by recording and repeating specific patterns of coordinated motion. If objects can be continuously conjoined enough times with the 'same' results, we say we know their 'cause'.

Abductive logic relies on a centripetal accumulation of empirical perceptions. Then it attempts to centrifugally wield a set of causal objects and practices back upon the accumulated periphery to control, direct and lead new and larger groups of objects.

Induction and hierarchy

The second logical operation of the ancient cardinal sciences was induction. Induction is the process of testing the consistency of various causal objects and hierarchically ranking them in order of more or less probable. Inductive reasoning, therefore, does not necessarily guarantee the accuracy of a prediction or causation. It merely establishes a hierarchy of cardinal objects based on how many other effect-objects someone can reliably coordinate with them. The more objects one can coordinate to a single cardinal object, the more powerful its inductive explanation – the larger its cardinality.

For example, from around the third millennium BCE onwards, both Mesopotamian and Egyptian civilisations began recording medical instructions to heal specific ailments and injuries. Over thousands of years, scribes continually amended these and other documents, such as engineering instructions and culinary recipes. Various hypotheses emerged, were collected through written inscription, tested by multiple people across multiple circumstances, and some were determined to work better or worse than others. Over time, the most potent cardinal objects of explanation were inscribed and centralised around the seat of political and religious power.

This was how an increasingly centrifugal inversion began. Inductive procedures such as laws and ritual practices were around for so long they seemed to be given from the gods. Kings received these records of best practices and applied them outwards and downwards on the masses.

However, an inductive hypothesis is not a divine, unchanging truth. It is only *found to be*, through trial and error, *more likely* than other hypotheses. Through centralised control over social and religious behaviours, ancient people could produce relatively consistent effects. This exposes three material and kinetic biases of induction.

First, inductive logic skews explanations towards the most visible and apparent objects. This tends to privilege objects that have already been selected and centrally accumulated around sites of epistemic authority. When a fetish object or rain talisman is dramatically visible during a successful rain prayer, and other objects, such as a distant and invisible approaching storm, are not, the talisman will be more likely to be coordinated to the rain.

Second, the fact that causal objects tend to emanate centrifugally from gods and kings in beer recipes, astronomical events and medical instructions also produces a confirmation bias. People tend to look for ways to confirm the current hypothesis emanating down and out from a central god or ruler, and tend to under-observe how it might not be valid. Errors in one view can be explained easily by competing gods or centres of causal power.

Finally, induction tends to look for simple explanations or a single cardinal object (god) for complex situations. However, there is no necessary reason why a simple answer is any more or less likely than a complex one. This bias was a political and centrifugal aspect of a centralised elite dominating the peripheral masses.

The more hierarchical and powerful the logical explanation, the fewer new hypotheses people consider. The more entrenched and centralised the hierarchy of inductive objects, the more stultified it tends to become. This is not unique to the ancient world. Kinetic patterns that persist over time shape inductive logic and imply a hierarchy and centralisation of causal objects over the periphery of effects.

Classification and reorganisation

The third stage of cardinal logic was classification. Classification implies both a division between ordered and unordered objects and a hierarchy of ordered objects. In addition to these previous operations, classification introduced a *reorganisation* of objects into new vertical and horizontal orders.

Classification is a fundamentally cardinal and centrifugal form of knowledge. It expands the process of one-to-many coordination to *all types of things*. For example, an object is a categorical object or 'type' of thing if one coordinates it kinetically to many other objects. Classification is not a universal or pre-given object but a historical and kinetic one that emerged, like cardinality itself, during the ancient period.

The classification process began as a lexical activity performed by urban scribal elites to help govern the periphery. Cardinal classification had two primary aspects. A class of objects is the same in one sense but different in others. Identity and difference, whole and part, coexist. A category coordinates one to the other. These classes of objects can then be coordinated to one another to create larger and smaller classes, like sorted stacks of papers.

For example, early Assyrian tablets from the city of Nineveh provide lexical lists of animals. Scribes placed the dog, lion and wolf in one category – the ox, sheep and goat in another. The dog family itself was divided into various races such as the domestic dog, the coursing dog, the small dog and the dog of Elan. Thus, dogs were divided hierarchically by their proximity to the central city (domestic vs. wild). However, scribes ordered birds by their speed and habitat: birds of rapid flight, sea-birds and marsh-birds. Assyrian scribes classified insects according to habit, that is, those that attack plants, animals, clothing or wood. They classified vegetables according to their usefulness.[22]

Through the late second millennium or Middle Assyrian period, these heterogeneous categories of objects proliferated alongside cuneiform writing.[23] As they spread, they also became longer and more homogeneous in their orders. Lexical lists also became increasingly concentrated in the central palaces and temples as lists of things' natural order. In this way, lexical lists 'came to play a role in the management of power and legitimation of a world empire'.[24]

Lexical classifications did not provide directly causal or logical explanations, but rather offered instructions for the proper *typological* reorganisation of beings. For example, scribes classified gods by their genetic primordiality or who came first. Early writers ordered humans by the importance and centrality of their activity and their proximity to the centre where the gods appeared. Political, priestly, scribal labour was at the centre, and craftsmen, slaves, women, children and animals were at the periphery.

Over time, classification became increasingly consolidated and reached its apex with early Greek philosophical monism. Early Greek philosophers believed that everything was a single 'type' of thing, such as water (Thales), air (Anaximenes) or fire (Heraclitus).

Deduction

The final and most abstract move in ancient logic was deduction. Deductive logic is, as Aristotle succinctly put it in his *Prior Analytics*, 'A *logos* in which, certain things having been supposed [*protasis*], something different from those supposed results of necessity [*ex anankês sumbainein*] because of their being so.'[25] By presupposing some initial true statements, one can arrive at some necessary consequences of these statements. Since the initial premises (*protasis*) are assumed to be true, the conclusions (*sumperasma*) do not need to be proven by their effects in the world. The findings are confirmed purely by definition.

Deductive logic inverts the other concrete logics by beginning not with qualities but as if true statements formally preceded the world of concrete objects completely. After thousands of years of concrete logic, deduction acts as if it preceded the logics that produced it.

Yet the etymological residue of deduction's kinetic origins remains in Aristotle's description of logic. For example, the Greek word πρότασις (*prótasis*) that Aristotle used for things that one presupposes comes from the Greek words πρό (*pró*) + τείνω (*teínō*, 'stretch'). The *prótasis* is, therefore, kinetically and historically related to the act of measurement and *rope stretching*. The related Greek word ὑποτείνουσα (*hupoteínousa*), meaning 'to stretch under', and from which we get the English word hypotenuse, also comes from the Egyptian word *harpedonaptae*, meaning 'surveyor' or 'rope-stretcher'. The logical premise comes from the movement of the foot (*ped*).

The Greek word *symperásmata*, which Aristotle used to describe the 'conclusion' of a deduction, also has kinetic roots in the words συμ- (*sum-*), 'together', + πέρασμα (*pérasma*), meaning 'passage, journey, or threading'. The conclusion of a logical deduction allows multiple measuring flows (*prótasis*) to flow or be threaded together (*symperásmata*).

The content of the flows that are threaded together, according to Aristotle, are the χορός, *horos* or 'terms'. These can be either universal (*katholou*) or particular (*kath' hekaston*). The Greek word *horos* comes from the Greek word χορα, *chora*, meaning both a space for dancing, made by the dance itself, as well as the gathering flow of nutrients from the countryside. A brief look at the original Greek meaning of these terms reveals that the material-kinetic origins of deduction (*sullogismos*) are based on a gathering (*lego*), measuring (*prótasis*) and threading together (*symperásmata*) of mobile objects (*choros*).

The 'threading together' of logic was also related to the Greek word for knowledge, *metis*, meaning shapeshifting, measurement and weaving. *Metis* comes from the Greek word μέτρον, *métron*, meaning measurement.[26] The origins of deductive logic are not pure ideas in the mind, but rather the movements of measurement, weaving and threading.[27]

The practice of logical deduction emerged at the geopolitical centre of a growing colonial empire with Thales' theorem, 'All triangles inscribed in a semi-circle will be right triangles.' Thales and Pythagoras travelled extensively in Egypt and Babylon and collected an immense diversity of inductive knowledge recorded by scribal elites. The Greeks did not invent science and mathematics, but rather removed the results of thousands of years of inductive logic from their geographical periphery in the Near East and synthesised it into a central and theorematic form of 'universal truth'.[28] A theorem is a statement that has been proven correct based on previously established premises. It is, therefore, fundamentally deductive.

However, the master-stroke of Greek theorematics was not that it was a mere synthesis of existing practical knowledge. The ancient Greeks went one step further by declaring that the theorem was ontologically primary, self-proving, and *creative* of all concrete knowledges. In classical Greece, Pythagoras, Philolaus, Plato and Euclid all made this move, which was only implicit in Thales. As Philolaus, a Pythagorean, was reported to have said, 'all things which can be known have number; for it is not possible that without number anything can be either conceived or known'. And further, that number was 'great, all-powerful and *all-producing*, the beginning and the guide of the divine as of the terrestrial life'.[29]

Ultimately, Euclid's celebrated *Elements of Geometry* provided the capstone synthesis of all Greek theorematic knowledge from Thales onwards. Euclid not only introduced definitions of elementary geometrical objects such as point, line and plane. From them, he deduced various axioms or postulates, which together proved without empirical demonstration an interconnected system of theorems. This included twenty-three definitions, five geometric postulates, and five additional postulates that he called 'common notions'. Using only these, Euclid was able to prove 465 theorems.[30]

Deductive knowledge was the knowledge of categories and their hierarchical relations of containment. That is all. The conclusion of a deduction is necessary because the definitions of the categorical terms used already contain it. Deduction creates no new knowledge of the world, only a knowledge of knowledge itself. It is the ultimate self-reflexive, circular and tautological movement, like the perfect rotation of a sphere in Aristotle's cosmology. The deductive centre remains eternal, perfect and immobile, while the world's periphery spins and changes around it. Deduction relies, like the centrifugal circle, on the self-sameness and identity of its cardinal whole. Many objects can be coordinated to a single object only if that single object remains what it is. This leads to the identity and oneness of being.[31]

Conclusion

This chapter concludes Part II on the ancient cardinal object and its centrifugal pattern of motion. However, the problem that emerged next in history was how these cardinal objects and their orders could be connected. This dilemma provoked the invention of a whole new kind of 'tensional' object that we will explore in the next section.

Notes

1. Norman Biggs, *Quite Right: The Story of Mathematics, Measurement, and Money* (Oxford: Oxford University Press, 2016), 23.
2. 'Only a material whose every sample possesses the same uniform quality can be an adequate form of appearance of value, that is a material embodiment of abstract and therefore equal human labour. On the other hand, since the difference between the magnitudes of value is purely quantitative, the money commodity must be capable of purely quantitative differentiation, it must therefore be divisible at will, and it must also be possible to assemble it again from its component parts. Gold and silver possess these properties by nature.' Karl Marx, *Capital Vol 1* (New York: Penguin, 1976), 184.
3. Biggs, *Quite Right*, 18.
4. André Leroi-Gourhan, *Gesture and Speech*, trans. Anna Bostock Berger (Cambridge, MA: MIT Press, 1993), 332.
5. Biggs, *Quite Right*, 25.
6. Hans-Jörg Nissen, Peter Damerow, Robert K. Englund and Paul Larsen, *Archaic Bookkeeping: Early Writing and Techniques of Economic Administration in the Ancient Near East* (Chicago: University of Chicago Press, 1993). Arithmetic (addition, subtraction, division) existed by 3100 BCE.
7. Biggs, *Quite Right*, 19.
8. See Edmund Husserl, *Philosophy of Arithmetic: Psychological and Logical Investigations with Supplementary Texts from 1887–1901*, trans. Dallas Willard (Dordrecht: Kluwer Academic, 2003), 193–5.
9. 'How I propose to improve upon it can be no more than indicated in the present work. With numbers . . . it is a matter of fixing the sense of an identity.' Gottlob Frege, *The Foundations of Arithmetic: A Logico-Mathematical Enquiry into the Concept of Number* (Evanston, IL: Northwestern University Press, 1999), x.

See also Bertrand Russell, *Principles of Mathematics* (1920) (London: Routledge, 2015), introduction.

10. 'Mathematically, addition is a fundamental operation. Subtraction is defined in terms of addition and cannot exist without it.' Lucas Bunt, *Historical Roots of Elementary Mathematics* (New York: Dover, 2012), 9.

11. 'In a similar manner the words "bi-partite", "tri-partite", etc., express the number of the parts of the whole, conceived of as an extrinsic property of that whole. All of these and similar concepts have an obviously secondary character.' Husserl, *Philosophy of Arithmetic*, 12.

12. Uta Merzbach and Carl Boyer, *A History of Mathematics* (Hoboken, NJ: Wiley, 2011), 12.

13. For more examples, see Merzbach and Boyer, *A History of Mathematics*, 12.

14. Biggs, *Quite Right*, 22.

15. Leonard Mlodinow, *Euclid's Window: The Story of Geometry from Parallel Lines to Hyperspace* (New York: The Free Press, 2001), 8.

16. David Burton, *The History of Mathematics: An Introduction* (Boston: Irwin McGraw-Hill, 2011), 47.

17. Cited in Petr Beckmann, *A History of Pi* (New York: Barnes and Noble, 1993), 23.

18. 'Observed facts relate exclusively to the particular circumstances that happened to exist when they were observed. They do not relate to any future occasions upon which we may be in doubt how we ought to act. They, therefore, do not, in themselves, contain any practical knowledge. Such knowledge must involve additions to the facts observed.' Such an addition, Peirce adds, 'may be called a hypothesis', or what he will also call an 'abductive inference'. Charles S. Peirce, *Philosophical Writings of Peirce* (New York: Dover, 2012), 150.

19. Charles S. Peirce, *Collected Papers of Charles Sanders Peirce, Volume V: Pragmatism and Pragmaticism*, ed. Charles Hartshorne and Paul Weiss (Cambridge, MA: Belknap Press of Harvard University Press, 1965), 171.

20. Peirce, *Collected Papers, Volume V*, 172.

21. *sSmt* is the Egyptian word for 'working out' or 'procedure', and also means governance, control. Many of these late lexical lists include dedication prayers, indicative that writing and education were closely associated with temples and political leadership. See William Brown, 'Cuneiform Lexical Lists'. *Ancient History Encyclopedia*, last modified 5 May 2016, available at <https://www.ancient.eu/article/900/> (last accessed 4 March 2021).

22. M. Civil, *Mesopotamian Lexicography* (Leiden: Brill, 2002).

23. N. Veldhuis, *History of the Cuneiform Lexical Tradition* (Münster: Ugarit-Verlag, 2014), 226.

24. Veldhuis, *History of the Cuneiform Lexical Tradition*, 391.

25. Aristotle, *Prior Analytics*, Book I, Section II, lines 24b18–20.

26. Here's how Beekes describes the etymology of *metis*: 'An original verbal noun meaning ★"measuring", *metis* is derived from the root ★*meh1-* found in Skt. *mimati* "measures", etc. The formation ★*meh1-ti-* itself is found in an isolated

Germanic word, OE *maed* [f.] "measure", and is presupposed by the denomi-
native Lat. *metior* "to measure".' Robert Beekes and Lucien Beek, *Etymological
Dictionary of Greek* (Leiden: Brill, 2010).

27. These terms are even explicitly entangled at the foundation of Plato's entire
theory of geometric forms, of which, he says, all reality is composed. In his
Timaeus, Plato says that before there were any geometrical shapes there was
first the *chora*, composed of nothing but moving flows [*dunamis*] of matter.
From these irrational [*alogon*] flows came geometrical forms. The Greek word
dunamis also means square root and is defined by Theaetetus in the dialogue
that bears his name (147D). Thus *dunamis* is both choric and the condition of
geometrical form at the same time: kinomorphism.

28. 'Thales of Miletus, hailed by Hellenophiles as the creator of science (see the
epigraphs at the head of this chapter), was reputed to have spent many years
abroad studying the ancient wisdom of Egyptians, Babylonians, and Phoenicians; it was even said by some that he was himself of Phoenician stock.
According to no less an authority than Aristotle, "the mathematical arts were
founded in Egypt". Plato attributed to Egyptian wisdom not only the invention of "arithmetic and calculation and geometry and astronomy", but also
"the discovery of the use of letters". "Geometry", Herodotus declared, "first
came to be known in Egypt, whence it passed into Greece." In the first century C.E., Strabo commented: Geometry was invented, it is said, from the
measurement of lands which is made necessary by the Nile when it confounds
the boundaries at the time of its overflows. This science, then, is believed to
have come to the Greeks from the Aegyptians; astronomy and arithmetic from
the Phoenicians; and at present by far the greatest store of knowledge in every
other branch of philosophy is to be had from these [Phoenician] cities [Sidon
and Tyre].' Clifford Conner, *A People's History of Science: Miners, Midwives, and
Low Mechanicks* (New York: Nation Books, 2009), 123.

29. Merzbach and Boyer, *A History of Mathematics*, 49, my italics.

30. Mlodinow, *Euclid's Window*, 34.

31. However, as we saw in the previous chapter in the case of the cardinal numeral,
the paradox of cardinal totality is precisely that the one that contains cannot be
included in that which it contains and thus remains something different from
itself.

III THE INTENSIVE OBJECT

9

The Tensional Object

Intensive objects are much stranger than cardinal objects. We usually do not think of them as distinct objects, yet we use them every time we count. In my definition, part of intensive objects is 'the action of counting' itself. Every time we order objects in a sequence before and after, or we count a whole series as 'one', we are *doing* something. It may be less intuitive, but that 'doing' is something that is part of the object.

Furthermore, while we are busy ordering and counting our objects, the world is also *doing something*. Nothing stands still. Objects are always moving and changing, and yet when we use ordinal and cardinal objects, we act as though everything is frozen and as though we, the coordinators of objects, are not made of moving objects.

What was so brilliant about the medieval and early modern sciences, from the fifth to the eighteenth centuries, was how seriously they took the invention and investigation of these strange but immanently practical objects. It was a novel idea to turn the study of abstract objects such as numerals *itself* into a new area of study. The science of intensive objects was also the first to take the process–nature of objects seriously. Although the aim of what I am calling the intensive sciences was ultimately to try and reduce movement to a new kind of object, they nonetheless grasped something extremely critical about objects and the counting of objects: movement and mutability.

The intensive sciences were sciences of change. They were interested in objects such as rates of change, degrees of change, qualitative changes, quantitative changes and non-finite numbers.[1] They wanted to know what exactly happened in the transition from one object to another.

The following chapters of this section on intensive objects aim to illustrate the kinetic patterns and processes that created and sustained these new intensive objects. This chapter argues that intensive objects emerged and were propagated in medieval and early modern Europe through three primary processes: tensional motion, deformation and polygonalism. Here, as in Chapters 4 and 6, I offer only the very broadest strokes of these processes. In the next two chapters, though, I describe how these processes emerged from various medieval and modern scientific practices.

What is an intensive object?

An intensive object is an object whose quantity is not fixed or finite but is variable. Unlike an ordinal or cardinal object, intensive objects are not 'this' or 'that' state, but a rate, degree or process by which one object transforms or transitions into another.

This is such a strange kind of object that many ancient and modern philosophers of science and mathematics refused to accept it. For example, Aristotle defined mathematics as concerned only with 'things which do not involve motion'.[2] Since intensive magnitudes vary and alternate, they cannot be proper mathematical objects for Aristotle. The Anglo-Irish philosopher George Berkeley even called intensive quantities 'objects of faith'[3] or 'evanescent increments'. They are, he said, 'neither finite quantities nor quantities infinitely small, nor yet nothing. May we not call them the ghosts of departed quantities?'[4] The German philosopher Georg Friedrich Hegel defined mathematics as the science of 'the quantitative form in which alone it considers [objects]'.[5] However, he wrote, since 'the infinitesimal calculus permits, even requires, procedures which mathematics must absolutely reject when operating with finite magnitudes, and at the same it treats its infinite magnitudes as if they were finite quanta . . .',[6] infinitesimal calculus must be rejected. There can be no science of intensive or fluctuating quantities for Hegel.

Intensive quantities are processes of variation, motion and alternation, such as the slope of a curve, the velocity of an object in motion, the infinitesimally small or the infinitely large. No single finite cardinal object can capture such objects.

Tensional motion

The first feature of this new science of intensive objects is what I call its 'tensional' motion. What is a tensional motion? A tensional motion is where linked objects move together relative to one another. For example, the human arm has a tensional motion because it has segments, hand, forearm and upper arm, held together by joints. The joints allow the parts of the arm to hold together and apart and swing and rotate around these joints. The springs and gears of a mechanical clock also have a tensional motion. When you wind up the spring, its tensional movement slowly releases through the tensionally linked gears. The gear-train of a bicycle works by the same motion. The crank arms pull a chain in tension with gears.

What does this have to do with the early modern science of objects? In this section, I hope to show that the invention and development of intensive objects relied on a distinctly tensional pattern of motion. For instance, one crucial and widespread technique of the modern era for studying the alteration of objects in time and space was the coordinate system. A coordinate system was like a map with latitude and longitude. Every position on the map had a horizontal and vertical value. As the position of the object changed, one could record each stage in two dimensions. The result was a visual history of the steps and degrees of change in an object relative to two intersecting axes.

I am calling this a 'tensional' motion because this new kind of object was not a single fixed state but a shifting proportion or ratio between two different coordinates in relative tension with one another. The measurer *held together* two changing aspects such as speed and distance in the same diagram, graph or map. This allowed scientists to identify regular correspondences between objects through time.

Medieval and early modern scientists also realised that by establishing a tension or coordinate system between changing objects, they could assign quantities to objects that no one had before quantified, such as temperature, colour and speed. However, this new method posed a novel challenge to quantification as well. If intensive objects are measures of change and transition, when do you stop measuring them? When are they done changing so that one can take the final measurement?

Many ancient thinkers were aware of intensive objects such as irrational numbers, π, Zeno's paradoxes and Archimedes' method of exhaustion. However, these objects did not become the basis of a whole new theory of objects. For the most part, the ancients treated them as interesting or paradoxical exceptions to the rules of cardinal objects.

The intensive sciences' fascinating discovery was that there was no 'final' or cardinal measurement. If intensive objects happen in the tensions *between* objects, then there is always more to measure as long as there is a difference between two objects. When are we done measuring a curve? We can always use one more tangent line to achieve a more precise measurement. When are we done counting π? There is always one more diameter through a circle to consider as a ratio of its circumference. With intensive objects, we make the path of measurement by walking.

To my mind, the most ingenious insight of medieval and early modern science was that all objects were *held* together by processes. 'Things', as I defined them in Part I, hold together qualities and quantities through iterative patterns of motion. When humans manipulate objects, they perform some action, even if it is the action of thought 'in their heads'. Actions and objects are not immaterial but occur in space and time, even when we *act as if* certain things have no qualities. These actions are tensions that hold groups of objects in a sequence or treat them 'as one'. As actions, these tensions can be measured, counted and manipulated as objects as well. The act of coordinating objects with our brains and bodies is an act of *holding* things together. From a kinetic perspective, this is how I am interpreting the science of intensive objects.

This is also why I say that *intensive* objects are 'tensional'. The Latin verb *teneo* means 'to hold', and from this root comes the Latin word *tendō*, 'I stretch', and from this *tensiō*, meaning 'tension'. Every counted object requires an act of counting to support it and connect it to others. By trying to quantify change and process as objects, the intensive sciences discovered the importance of the iterative act of following the changes and mapping the tensions between which aspects of the object remained stable, and which altered over time.

What then is 'infinity' from this material and kinetic perspective? Infinity, in my view, describes the way that objects are composed of non-finite and non-discrete *processes*. If things are always in motion and changing, then no single 'state' or finite cardinal can ever fully capture them *as processes*. In my interpretation, the word 'infinite' designates the negative capacity of objects, and all processes, to resist *discrete* enumeration.

However, because so many people have used 'infinity' as if it described a positive, transcendent or even immaterial aspect of objects, I hesitate to use this term without some qualification. In the next two chapters I will have occasion to explain my interpretation in more historical detail. For now, I will say that I do not believe that any object exists before the act that brings it into being. Even our thoughts of objects are themselves kinds of objects made through action and motion.[7]

What George Berkeley called the 'ghost' of quantity is, in my view, the material action of quantification itself that 'haunts' all objects.[8] The kinetic act that encloses and holds quantities together is not part of the group it encloses. We can count it as part of that group only by the action of counting, which will again escape the group we are counting. The real spectre of the sciences is, therefore, the practical process of creating and sustaining objects.

Platonist mathematicians have tried to exorcise this ghost by pretending that numbers exist independently of being counted. In this way, they constitutively exclude the act of counting from what it counts. However, the action of counting escapes all finite objects because it is the moving act that ties discrete quantities together and puts them in order.

Intensive science is also tensional insofar as it realised that the *act* of quantification *holds* qualities and quantities together in hands, pens and paper, charts and graphs, telescopes and microscopes, laboratories and manuscripts. The material fragility of knowledge and the importance of preserving it, translating it and transmitting it was something the medievals and early moderns learned after losing so much recorded knowledge in antiquity. For knowledge to persist, it must be physically held, bound together in tension, continually counted and recounted, coordinated and re-coordinated. Objects are always moving and changing, and so they have to be continuously held in place without end.

Deformation

The second feature of intensive objects is what I call their continual 'deformation'. By this, I mean that intensive objects are always altering their shape. We cannot study intensive objects as Plato and Aristotle once studied cardinal objects. For Plato and Aristotle, the forms were eternal and immutable. Objects were only different *in kind* and not *in degree*.

Intensive quantities are quantities in motion, rates of variation and continual alterations of quality corresponding to quantities. For the ancients, certain objects were fundamentally qualitative and could not be quantified, like the colour white, the good or motion. As the historian of science Carl Boyer notes, 'The Greek sciences of astronomy, optics, and statics had all been elaborated geometrically, but there was no such representation of the phenomena of change.'[9] The Greeks lacked a concept of acceleration and considered motion only with respect to uniform geometric patterns such as circular rotation. 'Motion was', Boyer tells us, 'a quality rather than a quantity; and there was among the ancients no systematic quantitative study of such qualities.'[10] Greek mathematics was the study of form rather than of variability.

All this began to change with the idea of *impetus*. Impetus is the notion that a body will continue to move once set in motion because of an internal tendency. By contrast, Aristotle thought that things continued to move because something like air continually impelled them from outside. The sixth-century Alexandrian philologist John Philoponus was the first to describe the idea of impetus and contrast it with Aristotle's dynamics. Unfortunately, Philoponus's work was mostly unread for centuries until it was rediscovered in the early fourteenth century by the French scientist Jean Buridan, who spread the idea of impetus throughout Europe.[11]

If objects are the source of their motion and alteration, their forms are *internally dynamic*, not fixed, eternal or immutable. In the thirteenth century, scientists described a similar idea called 'the latitude of forms'. The latitude of forms was a way to rethink objects as 'continuous magnitudes', like measuring sticks, along which one could measure degrees of change or 'continua'. Medieval thinkers called these degrees of fluctuation in objects *intensio* as they became 'more' and *remissio* as they became 'less'. This same idea of self-motion also played a significant role in creating automata or self-moving machines and mechanisms in the middle ages.

These ideas and others that I will discuss in the next two chapters were all part of a common shift in the pattern of motion of objects. Ordinal and cardinal objects persisted, of course, but they were also taken up and transformed by these new intensive objects. Instead of treating objects as merely fixed and finite centres of measurement, scientists began treating them as subject to all kinds of deformations and alterations. Intensive objects have forms that undergo continuous motion and deformation each time one measures, translates, reproduces or transmits them. This was what the 'Oxford calculators', a famous group of fourteenth-century thinkers, called *latitudo difformis*.[12]

However, the idea of the continuous deformation of objects was only possible, in my view, when scientists realised that objects were something formed by processes of internal tension, which changed in direct proportion with one another in motion. Unlike cardinal objects, there was no 'change in general' for intensive objects. Instead, there were rates of change.

A ratio is a kinetic tension between two dimensions of an object that varies continuously. Qualities can vary with quantities and quantities can vary alongside other quantities. In this sense, intensive objects have a unique 'many-to-many' coordination. Intensive objects are changes in the rates and ratios of several continuous and indefinite deformations.

Polygonalism

The third and final feature of intensive objects is their 'polygonalism'. By this, I mean that intensive objects are held together by rigid or inelastic links like the straight lines of polygonal shapes. The relationships between intensive objects are like a polygon whose straight lines intersect at movable joints. Such a shape can move but only relative to a fixed set of links among the lines.

This image helps me think about how the intensive sciences organised the objects of their world. They understood that objects were formed by moving and changing processes, but they also wanted to quantify those movements and changes. So they created maps and coordinate systems that could treat these processes as objects. Ultimately, however, in my view, processes cannot be reduced to objects, but that does not mean that nothing is gained from *acting* as though they can be.

Another helpful image for understanding how tensional objects work 'polygonally' is called the 'method of exhaustion'. One way to approximate a circle's area is to fill it with polygons, whose areas can be calculated finitely. How many polygons fit in a circle? An indefinite amount. As they approach the circle's perimeter, the polygons become 'arbitrarily small', without ever genuinely exhausting the circle. What I am calling 'polygonalism' is the indefinite process of quantifying an object that cannot be entirely or finitely quantified. Instead of dealing with single finite objects, the intensive sciences describe an endless series of finite stages of objects like a sequence of snapshot-objects run together to reveal a process.

However, since movement is not fundamentally discrete, there is always something more to quantify, count and coordinate. The intensive objects keep going as polygons keep filling a circle or as tangents keep intersecting a curve. There is no end to the series of intensive objects because we cannot reduce the objects' movement to a series of discrete static states. Objects exist in woven networks of relations, but these relations are always shifting as they move. The intensive sciences study these relations by capturing them into ratios of change. However, since relations are continually changing, the ratios can be endlessly approximated and readjusted.

I don't mean that all intensive objects are actually polygons. It is just an image to help show the general pattern at work. The intensive sciences create objects like the sides of a polygonal that are continually being added to, subtracted from and modified to approximate an object's irregular movement. The intensive field of objects is one where objects kinetically relate to one another through a polygonal network of tensions, latitudes, longitudes, ratios, proportions and rates of relational change.

Conclusion

This chapter has tried to sketch the general kinetic features of what I am calling intensive objects. In the next two chapters, I would like to show how these three features emerged historically in four significant sciences: kinematics, dynamics, the mathematics of infinity and the method of experimentation. Although these are very different scientific practices, my argument in this section of the book is that all four follow the same 'tensional' pattern of motion in their creation and use of objects.

Notes

1. Infinite ordinals and infinite cardinals would fall under my definition of intensive numbers because they are not finite and require a different pattern of motion.

2. Cited in Carl Boyer, *The History of the Calculus and its Conceptual Development* (1949) (New York: Dover, 1959), 72.

3. George Berkeley, *The Analyst: A Discourse Addressed to an Infidel Mathematician*, Sect. 7, in George Berkeley, *The Works of George Berkeley Bishop of Cloyne* (Nendeln: Kraus Reprint, 1979).

4. Berkeley, *The Analyst*, Sect. 35.

5. 'Pure mathematics, too, has its method suited to its abstract objects and the quantitative form in which alone it considers them.' Georg W. F. Hegel, *The Science of Logic* (Cambridge: Cambridge University Press, 2015), 32.

6. Hegel, *The Science of Logic*, 205.

7. This is another way of saying that I do not and cannot accept the axiom of choice as a basis for the philosophy of motion. See Chapter 14 for more on the axiom of choice.

8. This is what Cantor will later call an 'uncountable infinity'. See Chapter 14.

9. Boyer, *History of the Calculus*, 71.

10. Boyer, *History of the Calculus*, 72.

11. See Thomas Nail, *Being and Motion* (Oxford: Oxford University Press, 2018), ch. 25.

12. Boyer, *History of the Calculus*, 74.

10

The Medieval Object I

We often think that not much happened during the 'middle' or so-called 'dark' ages, but medieval scientists helped invent a whole new world of objects. No one before had ever dreamed of quantifying speed, movement or colour, and yet today it seems natural because of what happened during these centuries. The early intensive sciences built on ancient knowledge and broke with it on decisive issues that shaped the modern scientific revolution and our contemporary understanding of objects. Historians often lionise Newton as a genius of modern science, but much of what Newton wrote was built on or synthesised from work done by scientists in the middle ages.

In the next two historical chapters, I want to show how intensive objects emerged through four crucial scientific practices during this time: kinematics, dynamics, the mathematics of infinity and experimental method. My argument in these chapters is that all four of these sciences exhibited tensional patterns of motion.

This chapter begins by looking at two sciences of moving objects: kinematics and dynamics. Kinematics studies the forms of motion, while dynamics studies the causes of motion. Both wonderfully describe the variable and tensional nature of these new objects.

Kinematics

Kinematics is the study of the patterns, trajectories and rates of change of moving objects. However, before it was officially termed 'kinematics' by André-Marie Ampère in his *Essai sur la Philosophie des Sciences* (1834), it was known as the study of 'the latitude of forms'. Scholars used the term 'latitude' to describe the range of degrees a form took as it changed over time. Throughout the middle ages, there were several theories of these latitudes. In what follows, I want to trace the historical emergence of these theories so you can see how they became increasingly expansive in their scope of objects and precise in their tensional motions.

The latitude of forms

The second-century Greek physician Galen first introduced the idea of latitude or gradual alteration of forms in Book II of his *Microtegni*. He invented this term to describe the gradual continuity between states of health and illness. The Greek word Galen used for latitude was πλάτος, *platos*, meaning 'plane', 'surface' or 'breadth', because he diagrammed health as a continuous surface. He located three areas on this diagram. On one side was the state of optimal health, on the other side was the state of serious illness, and in the middle was a neutral state.[1] Instead of thinking about health as a discrete arithmetical state, he thought about it as a continuous geometrical state of changing degrees.

As novel as it was, Galen's idea of latitude was still limited to the changes of a single *subject*. By that, I mean that the changes were only changes in a subject affected to a degree by forms. For example, the subject can be more or less healthy, but the form of the illness itself does not change. Galen and those who followed him were still reluctant to admit that the forms themselves underwent deformation. This same limitation was true of the 'admixture theory' of Avicenna,[2] the 'actualisation theory' of Thomas Aquinas[3] and the 'succession theory' of Godfrey of Fontaines.[4] For almost a thousand years latitude theories persisted, but they remained solidly restricted to degrees of change in the affected subject. However, all this changed in the thirteenth century with the emergence of the 'addition theory' of latitude.

The addition theory

The addition theory of latitude was the first to concede that the latitude of forms applied *to the forms themselves*. This theory was defended early on by Henry of Ghent and the Franciscan theologian Richard of Middleton. Middleton's work then influenced John Duns Scotus, William of Ockham, Peter Auriol, Gregory of Rimini and the Oxford calculators, Thomas Bradwardine, William Heytesbury, Richard Swineshead and John Dumbleton.

According to the addition theory, quantities and qualities have *their own latitudes* within which continuous degrees can be distinguished 'in virtue of their nature and essence', as Henry of Ghent wrote.[5] 'As we can see', he said, 'there is no difference with respect to this topic between the magnitude of a body and the degree of a perfection.'[6] By 'magnitude', Ghent understood quantity, and by 'perfection', he meant quality. For Ghent, the qualitative changes of a form can be added to one another in series, just like changes in quantity. Both qualitative and quantitative series can be added to indefinitely. 'If the perfecting of a form, of which we have spoken, can proceed to infinity, then any increase by addition, considered absolute and *simpliciter*, can proceed to infinity.'[7] Middleton similarly argued that an increase in force, which previous scholars thought of as a quality, could be increased like a quantity of mass by addition.[8]

Scotus added to these ideas that degrees of quality were like the latitudes of a ruler. Each degree was a sum of all lower degrees and, therefore, *additive*. The whiteness of a wall, for example, Scotus wrote, could be continuously increased without changing the quantitative aspects of the wall. Or as Ockham put it, 'there are increases in intensity which are accomplished by the addition of one part to another; the second part constitutes a single thing with the first, but it is not distinguished from the first by its place and location (thus it is when a body which is completely white becomes more white than it was)'.[9]

In the fourteenth century, two of the Oxford calculators, Dumbleton and Swineshead, introduced the idea of a system of 'measurement' for qualitative changes. They argued that one could measure qualitative degrees of change and quantitative degrees of change *on the same continuous latitude*. They imagined this would be possible if they treated all qualities in the same way as they they mapped the movement or speed of objects with their quantitative distance. By coordinating an object's change in position with its *rate of change between positions* (speed) simultaneously, they put forward the first formulation of the mean speed theorem. This theorem stated that a body moving with constant velocity travels the same distance as an accelerated body in the same time if its velocity is half the final speed of the accelerated body.

Their new system of measurement distinguished between uniform rates of speed (*latitudo uniformis*) and non–uniform rates of speed (*latitudo difformis*). This method also led them to map other qualitative changes such as illumination, heat and density in a similar way.[10] Contra Aristotle, they argued that the rate of qualitative and quantitative change could be determined at any given instant. This is what they called an object's 'instantaneous velocity'. As Swineshead wrote, 'To every degree of velocity (i.e., qualitative or instantaneous velocity), there corresponds a lineal distance which would be described, assuming a movement throughout the time at this degree.'[11]

For all the Oxford calculators, a latitude was a homogeneous continuum, which one could present as a line on which the only differences were differences in length. According to the American historian of science Edith Sylla, this theory provided

> a better physical basis for quantification of qualities because the latitude corresponds to the intensity or degree of a quality at a point of

the body or in an instant of time and not only to some variation of the quality over its extension or over time. The latitude of velocity is imagined as a line. Equal parts of the latitude of velocity correspond to equal differences of velocity.[12]

The geometrical theory

The final and most precise theory of the latitude of forms was known as the 'geometrical theory', because it graphed the simultaneous changes in the qualities and quantities of objects as geometrical shapes. Historians sometimes attribute this to the medieval French theologian Nicole Oresme. However, according to the American historian Marshall Clagett, it was an Italian Franciscan named Giovanni Casali who first described the basics of the idea in his book, *On the Velocity of the Motion of Alteration* (*De velocitate motus alterationis*) in 1346. The geometrical theory was also implicitly suggested by Swineshead, who compared the geometrical length and breadth in an area with the lines that recorded quantitative change, which he called 'extension', and qualitative change, which he called 'intension'.[13]

In his book, *On the Configurations of Qualities and Motions* (*De configurationibus qualitatum et motuum*) (c. 1350s), Oresme also cites as precursors to his idea that Aristotle used lines to represent periods of time; that Grosseteste used lines to imagine the 'intension of light'; and that the Euclid commentator Giovanni Campani said that one could imagine anything continuous by a line, surface or body.[14] Oresme drew on these ideas using the geometrical coordinates of latitude and longitude to map the area and rate of intensive and extensive changes of objects.

Oresme's work was also vastly superior in detail and clarity to that of any of his precursors. The German historian of science Anneliese Maier characterises 'Oresme's method of graphical representation' as 'undoubtedly the most original achievement of the fourteenth century'.[15] His great innovation was to treat all changes in objects as *continuous kinetic changes in two geometrical coordinates*.

Oresme invented what we might call an 'intensive geometry' from three related historical ideas. First, he took Aristotle's notion that alteration always involves a continuous motion. Aristotle called this a 'pure quality'. Then, Oresme drew on Euclid's belief that one could imagine all continuous objects as geometrical objects. Finally, Oresme borrowed the Oxford calculators' notion that one could imagine all qualitative changes as continuous geometrical lines. Putting these together, Oresme brilliantly surmised that insofar as 'every measurable thing is a continuous quantity', it must also have continuously changing qualities. He imagined qualitative and quantitative changes as two straight lines tracing a geometric shape in their relationships through time.

Oresme treated these new objects, or what he called 'intensible things' (*rei intensibilis*), as geometrical surfaces or areas composed of two continuously changing dimensions. The first dimension traced the qualitative changes of an object. He called this qualitative dimension of change 'intensive change', since it described how an object became more or less of a quality such as whiteness or fastness. Intensive changes, such as changes in colour, for Oresme, were not additive because they did not have parts. For example, if you paint a white fence white, it does not become more white. However, as the fence gathers dirt, it can become *relatively* 'less white'. In section I, I called these kinds of qualitatively ranked objects 'ordinal objects'. These are the same kinds of objects Oresme had in mind when he described 'something which is said to be "more such and such" "more white" or "more swift"'.[16]

The second dimension of Oresme's geometric diagrams plotted the quantitative changes of an object. He called this quantitative dimension of change 'extensive change', since it described how objects gained or lost parts in space and time. Extensive modifications, such as stacking two one-inch blocks on top of one another to create a two-inch-high block, are additive. These are the kinds of objects I called cardinal objects in the previous chapters.

Oresme mapped each kind of sequence along its geometrical line. He drew intensive changes along a line of latitude and extensive changes along a line of longitude.[17] He then plotted as a third line the ratio of change

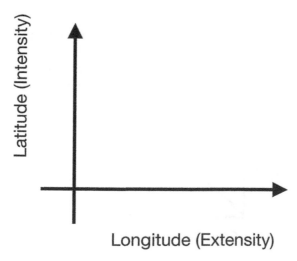

Figure 10.1 Extensive/intensive axis. Author's drawing.

between each. Mapping the rate of change in this ratio of coordinated suc–
cessions draws a geometrical figure or area, for example a square or triangle.

Using this method, Oresme could calculate the distance travelled by an
object by mapping the ratio of its changes in time as a longitude with its
changes in speed as a latitude. If the rate of change remained the same (*uni–
formis*) the resulting figure looked like a square. If the rate of change changed
at a constant rate (*uniformis difformis*), it looked like a triangle. If the rate of
change changed at a variable rate, it looked more like a curve or irregular
polygon (*uniformis difformis difformis*). One could then quantitatively measure
the *areas* produced by these shapes following the rules of geometry.

However, Oresme's geometrical method should not be confused with
the Cartesian coordinate system, since it did not plot 'points' on a grid, but
instead drew lines into geometric areas.[18] Oresme's geometrical method's
importance and novelty were that it maintained the tension between heterogeneous qualitative and quantitative changes. His diagrams did not unify
the two in a series of points but traced their trajectories into shapes.

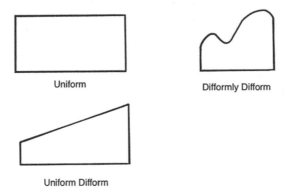

Figure 10.2 Uniform, uniformly difform and difformly difform. Author's drawing.

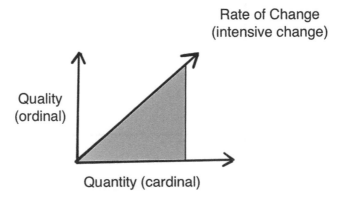

Figure 10.3 Oresme's theory of intensive change. Author's drawing.

Intensive objects

Now that I have charted the historical development of the study of latitude, I want to conclude this discussion by clarifying how medieval kinematics relates to my concept of 'intensive objects'. I also want to spell out
how kinematics followed a tensional pattern of motion.

What I call an 'intensive object' is different from what Oresme and others called 'intensive' or 'non-additive qualitative' changes in objects. In my
usage, an 'intensive object' is neither the intensive-ordinal nor extensive-
cardinal object alone, but the ratio of changing series of such objects graphed
together simultaneously on the same plane. The *ratio itself* is the new third
intensive object made from the coordinated tension between ordinal and
cardinal series. With intensive objects, one can no longer measure an object
once and for all but must enter into a *continuous measurement* process that
tracks the change and deformation of things.

What came to the foreground of knowledge during the middle ages was
the tensional relationship or ratio *between* changing ordinal qualities and cardinal quantities discovered in the act of measurement. 'Measurement', in the
middle ages, as the French philosopher and mathematician Gilles Chatelet
wrote, 'must above all be understood as an *act of knowledge*, as an articulation
of a degree of intelligibility – here a more or less swift conquest of space by
thought – and a rhythm of juxtaposition of parts extracted from a whole.'[19]

The intensive object, considered on its own, was a purely relative object
made by process of motion, or 'pure alteration', as Aristotle called it. The
unit of measure, the act of measurement and the measure's object were not
separate but instead were treated as three dimensions of a single intensive
object. For the intensive sciences, the *activity* of the object itself became a
new object of study.

There are two main tensional aspects of these objects. The first is the
tension between the qualitative and quantitive aspects of objects. In Chapter 2, I said that a 'thing' is a metastable process and has a qualitative and
quantitative aspect. One of the limitations of abstract cardinal objects was
that they treated objects as if they had no qualities. The intensive sciences
are so fascinating because they tried to hold the changing qualities and
quantities of objects together for a more rounded knowledge of objects.
This first tension between qualities and quantity is not only a real material
tension in objects themselves but something the sciences physically strug-
gled to hold together using new graphing and charting techniques.

The second kinetic tension held together a whole sequence of objects
by graphing each with changing qualitative and quantitative dimensions.

Intensive objects are held in tension, intensively and extensively. For
things to persist, they need to maintain relatively stable ratios between their

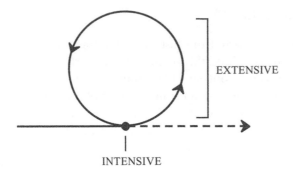

Figure 10.4 Intensive and extensive aspects of the fold. Author's drawing.

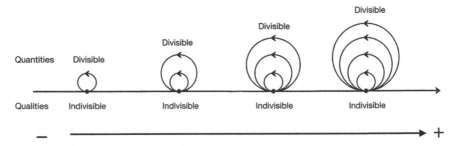

Figure 10.5 Series of intensive and extensive folds. Author's drawing.

changing qualities and quantities. If not, they fall apart.[20] The medieval scientists of intensive objects understood this. Their work laid the foundations for later thinkers such as Galileo, Buridan, Copernicus and Newton, as I will show in the next section.[21]

Dynamics

The second intensive science during this time was dynamics. Dynamics is the study of the *causes of change* in moving objects. Scientists called these causes 'forces', such as 'impetus', 'inertia', *'conatus'* and *'vis impressa'*. These forces described the proportional changes of acting and reacting bodies in motion. Where kinematics had focused on the *rates of change* within a single body, dynamics focused on the causes of bodies' collective movement.

My argument in this section is that the dynamic sciences emerged out of the study of kinematics and relied on the same tensional pattern of motion. Dynamics also expanded the study of intensive objects beyond the theoretical and graphing models of kinematics to the experimental sciences of physics, astronomy, optics, biology, chronometry, hydraulics and magnetism. These new sciences also required new measurement technologies such as the pendulum, telescope, optical lens, microscope, clock, vacuum pump and compass. I want to show that these technologies also followed the tensional pattern of motion that shaped intensive objects.

Terrestrial dynamics

One of the earliest areas of dynamics was the study of the causes of earthly movements. Why do things keep moving after something sets them in motion? Aristotle imagined that little puffs of air propelled a thrown javelin forwards, just as he imagined that god, or what he called the 'unmoved mover', continually pushed all things along. This was an ancient and centrifugal understanding of causality and motion in which a single static centre caused the motion of the periphery.

However, throughout the medieval and early modern period, a new understanding of the cause of motion emerged. It was called 'force', or *rhope* in Greek, by the sixth-century Christian theologian John Philoponus. For Philoponus, a force was not unidirectional or homogeneous, but rather a rate of change inversely proportional to the friction of its medium. The French priest Jean Buridan eventually popularised the idea of force across Europe.[22] However, it was not until the work of Galileo Galilei and his original study of the pendulum around 1602 that the concept of force, or what he called *impetus*, had a practical application in dynamical science.

The pendulum

The pendulum demonstrated the force of things concretely and facilitated its measurement. According to his biographer and student, Vincenzo Viviani, Galileo discovered the regularity of a pendulum's motion by coordinating his pulse with the swinging motion of a chandelier in Pisa Cathedral.[23] Galileo was the first to observe that a swinging weight around a single pivot takes the same time to get from one side to another despite the variations in how far it moves. The pendulum showed that the *acceleration* rate between oscillations slowly decreases proportionally to the decrease in the amplitude.

Galileo also argued that the period of the swing is approximately related to the length of the string. The time it takes for the pendulum to oscillate once, squared, equals the string's length. Even more impressively, Galileo extrapolated from this that without air or friction, the initial 'heavy falling body acquires sufficient impetus to carry it back to an equal height'.[24]

The kinetic pattern of the pendulum was tensional. There was a tension in the string between the pivot point and the swinging weight. There was also tension between the motion of the weight and the friction slowing it down. Since the weight does not return to the precise point where one released it, Galileo reasoned that some other force must be acting in tension against the moving weight, slowing it down. The great thing about the pendulum was that it was a reproducible mechanical set-up that one could use to measure force relatively accurately.

The pendulum rendered visible and predictable, for the first time, the tensional and proportional relationship between the force of movement and the counter-force of inertia. It helped create a new dynamic object composed of a moving tension or correlation between the indivisible or 'intensive changes' of speed and the divisible 'extensive' number of oscillations. In this way, the discovery of the dynamic object of force was deeply bound up with the pendulum's materiality and kinetics. Without it, scientists only had access to the more theoretical graphs and charts of Oresme. The pendulum was one of the first instruments to experimentally measure dynamic quantities such as the rate of change in acceleration, amplitude and periodicity in a moving object.

Celestial dynamics

Another crucial intensive science was celestial dynamics. Celestial dynamics studied the causal forces of heavenly bodies, just as terrestrial dynamics studied the forces of earthbound bodies. Throughout the rise of modern European science, the ancient geocentric model of concentric spheres and unchanging forms increasingly fell out of favour. Scientists wanted to know what caused the heavens to move the way they did. Modern astronomers introduced the idea that the motions of celestial bodies were held together and apart by the *same* tensional forces of attraction and repulsion that were at work on the earth. This was an enormous change from the ancient view of the heavens as following completely different laws.[25] Instead of a radical difference between heaven and earth, modern scientists saw a tension between two aspects of the same physical world.

The telescope

The invention of the telescope contributed immensely to Copernicus's heliocentric hypothesis. It made visible the dynamic and tensional motions of the planets, moons and comets for the first time. The telescope itself also worked by tensional movements.

We can, therefore, identify three kinetic stages of celestial observation. Prehistoric observation followed a centripetal pattern as light gathered from the sky's periphery to the centre of the eye. Ancient sundials followed a centrifugal motion by creating a fixed object that projected its shadow on to a peripheral surface. Modern telescopes had a new tensional motion that used concave and convex lenses to refract, contract and bend light flow to a focused point. This made something large look small, and something distant appear close. By combining a convex lens on one side of a tube and a concave one on the other, the telescope could capture light and create an optical tension between contracted (convex) and expanded (concave) waves of light.

This optical tension also created several visual aberrations and light diffractions due to imperfections in the lenses and the process of focusing divergent light waves. Where diffracted waves of light converge, one can see the celestial object in focus. Where they diverge, one sees rings and auras of light. In this way, telescopic diffraction made possible magnification and distortion through the tension of light waves.

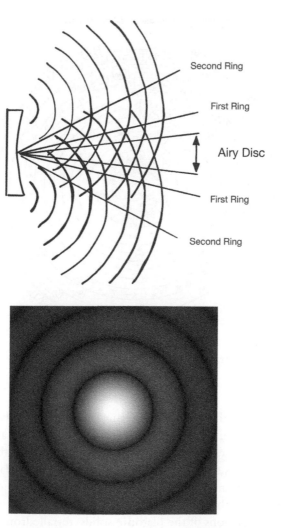

Figure 10.6 Telescope diffraction and airy disc. Top: author's drawing. Bottom: https://en.wikipedia.org/wiki/Airy_disk#/media/File:Airy-pattern.svg.

Like the pendulum, the telescope was a tensional object that made it possible to coordinate qualitative changes in celestial objects with their quantitative changes in space and time. The result was a whole new intensive science of astronomy.

Celestial bodies

Kepler, Galileo and Newton all used the telescope to support the idea that the earth rotated on its axis. Instead of perfect concentric circles, these astronomers imagined more triangulated and polygonal relationships between celestial bodies.

Figure 10.7 Concentric universe. https://en.wikipedia.org/wiki/Celestial_spheres#/media/File:Ptolemaicsystem-small.png.

It is no coincidence that, for Kepler, the celestial triangulation of forces made possible a 'divine ratio' or 'golden triangle' in mathematics, geometry and all being.[26] For Kepler, the language of nature was mathematical, and the name of God was the divine polygon. God held the planets together, but his force was plural and indirect in a fluid medium of aether. For Kepler, the orbits of the planets were held in tension with one another in the shape of the polygons of the platonic solids: tetrahedron, cube, octahedron, dodecahedron and icosahedron.

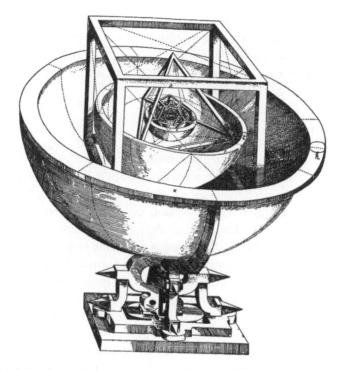

Figure 10.8 Kepler's polygonal universe. https://en.wikipedia.org/wiki/Johannes_Kepler#/media/File:Kepler-solar-system-1.png.

For Kepler, celestial motion was no longer concentric but a dynamic and harmonic ratio of geometric shapes made by God, the geometer. The relations of attraction and repulsion between celestial objects took the form of precise proportional ratios, just like Galileo's laws of motion for pendulums.

Gravity

Newton called these tensional forces of attraction and repulsion among celestial objects 'gravity'. In his *Principia*, Newton claimed that the entire task of science was to investigate the phenomena of movement and determine their causal forces. Once scientists discovered the ratios and proportions of changing objects, others could apply these laws to all other motions.[27]

For Newton, causal force had a dual function. When a body is at rest, it has an inherent force (*vis insita*) that resists any body that 'endeavors to change its condition'. When it is in motion, it has the power to impress and transfer force externally to other bodies (*vis impressa*). 'It is resistance so far as the body, for maintaining its present state, opposes the force impressed; it is impulse so far as the body, by not easily giving way to the impressed force of another, endeavors to change the state of that other.'[28]

Gravity, for Newton, was a tensional force like a fluid that could be transmitted between objects, holding them together and apart at the same time. Gravity was 'the absolute force to the centre, as endued with some cause, without which those motive forces would not be propagated through the space round about'.[29]

Fluid dynamics

A third crucial intensive science was fluid dynamics. Fluid dynamics studies objects such as air and water according to their intensive and extensive rates of change. Ancient hydraulics had focused on moving fluids from a central source to a peripheral destination via canals, aqueducts, irrigation, dams, wells and cisterns. However, modern fluid dynamics focused instead on the changing velocities of fluid bodies with one another.

Fluid dynamics was less interested in the formal differences between squares and circles than in the continuous transition of square shapes into circular ones. Modern fluid dynamics used technical innovations such as the microscope and thermoscope to study blood circulation through the body and the velocity of water through pipes. Within fluid bodies, there is a tension between attractive and repulsive forces called 'pressure' and 'viscosity'. Let's look at two examples of these tensional forces, one from biology and one from hydrology.

The microscope

In 1628 William Harvey published *De motu cordis* (*On the Motion of the Heart and Blood*), in which he theorised that blood moved from the heart through the entire body, returned, and circulated. In 1661 the invention of the microscope confirmed the existence of capillaries and Harvey's thesis.[30]

The kinetic power of the microscope was similar to that of the telescope. Both used optical tensions of refracted light, and both allowed scientists to observe bodies at higher magnification *in motu*. In Galen's time, scientists gained anatomical knowledge by dissecting unmoving dead bodies. However, the microscope made it possible to watch the dynamic tensions between moving bodies in action. Scientists could then establish ratios and proportions of qualitative and quantitive changes in these bodies.

For example, using only a small hand lens, Harvey put together a whole new tensional theory of blood pressure and circulation. By contrast, Galen held a more centrifugal theory in which blood was created in a central location, the liver, and then distributed uni-directionally to the peripheral limbs. There the body would consume it. For Galen, the heart and lungs only regulated the body's temperature and were not related to blood circulation.

In Harvey's model, however, there was no longer a single centrifugal origin of blood. Instead, there was dynamic blood flow, pushed and pulled by pressurised tensions in arteries of different sizes. As blood moved along the arterial walls, it produced viscosities, frictions and pressures. He thus saw blood movement as dynamic and responsive to changes in pressure in the whole circulatory system.

Harvey was even able to measure the flow of blood by holding together two changing series of objects. He estimated that the heart released one-eighth of its blood capacity with each beat and that the heart beats about a thousand times a day. This blood volume multiplied by this heart rate equals more blood than is contained in the human body. Therefore, Harvey reasoned, blood cannot be created in the liver each day, but the body must circulate it.[31]

Thermodynamics

The fourth and final dynamic science I want to look at is thermodynamics. Thermodynamics is the study of heat and is related to fluid dynamics because it also studies the pressures, tensions and rates of change in fluid materials. Thermodynamics also follows the same tensional pattern of motion in its coordination of qualitative and quantitative changes using modern instruments. One of the key tools of this new science was the thermometer.

Thermometer

The thermometer and barometer allowed scientists to coordinate two sequences of changing objects into a single new intensive object. They observed the qualitative change of a fluid inside a pressurised tube and coordinated it to a series of equally spaced markings on the tube. The coordinated tension between these two series made possible a new intensive object of proportional change.

The earliest expression of this instrument was the thermoscope built by Galileo in 1593. Galileo attached a thin, vertically rising pipe with a large empty glass ball at the top to a small vase filled with water. As the upper ball's temperature changed, it would change the pressure on the water level below, causing it to move up or down in the vertical pipe.[32] The hotter the ball got, the lower the tension or pressure pushing down on the fluid, and the higher the fluid would rise in the tube.

In 1611 Galileo's student, Santorio Santorio, added a series of marks on the tube's side to quantify the movement. With this addition, Galileo and Santorio invented a new intensive object that coordinated the qualitative change of fluid to the marks' quantitative units. Changes in pressure, temperature and volume could then be linked together as inelastic proportional dimensions of change inside a single thermodynamic object.

Boyle's law

The thermometer and the barometer made it possible for scientists to conduct new experiments on the controlled rate of change in fluids. The temperatures of fluids and gases were interesting because their changes were so proportional and connected to the atmosphere more broadly. It was not enough to treat objects as unchanging wholes in isolation from their environment.

One of the most brilliant syntheses of this idea was Boyle's law. The Anglo-Irish natural philosopher Robert Boyle discovered that for a fixed amount of an ideal gas kept at a fixed temperature, pressure and volume were inversely proportional. The more pressure was exerted on a fixed amount of confined fluid at a fixed temperature, the smaller the fluid volume. In this way, Boyle showed a new intensive object composed of simultaneous proportional changes in temperature, volume and pressure in a continuously deforming object. The tension between these proportions makes this an intensive object.

Conclusion

This chapter has shown how some of the first intensive objects emerged through kinematics and dynamics. My argument was that these sciences of moving objects measured changes proportionally along two or more coordinated dimensions. This is why they followed the tensional pattern of motion I described in the previous chapter.

This chapter is not a complete history of intensive objects. Instead, its purpose was to give the reader a sense of how medieval and early modern objects differed from ancient ones in their pattern of motion and their measuring instruments.

Kinematics and dynamics were only two significant sciences at the heart of the invention of intensive objects. Two more sciences played critical roles in forming these new objects that I want to show: the mathematics of infinity and the creation of experimental methodology. These are the subjects of our next and final chapter in this section of the book on intensive objects.

Notes

1. Edith Sylla, 'Medieval Concepts of the Latitude of Forms: The Oxford Calculators', *Archives d'histoire doctrinale et littéraire du moyen-age*, 40.251 (1973), 227–8. A similar thesis is also confirmed by Marshall Clagett, *Giovanni Marliani and Late Medieval Physics* (New York: AMS Press, 1955), 34–6.
2. 'I do not like the middle position and, in fact, detest its claim that blackening is a quantity and augmentation a quality. It is not right that blackening is a

blackness that is undergoing intensification; rather, it is an intensification of its subject with respect to its blackness.' Avicenna, *The Healing* (Provo: Brigham Young University Press, 2005), Book II, ch. II, 130.

3. Mary Beth Mader, 'Whence Intensity? Deleuze and the Revival of a Concept', in Alain Beaulieu, Julia Sushytska and Edward Kazarian (eds), *Gilles Deleuze and Metaphysics* (Lanham, MD: Lexington Books, 2014), 234.

4. Jean-Luc Solère, 'D'un commentaire l'autre: l'interaction entre philosophie et théologie au Moyen Age, dans le problème de l'intensification des formes', in M.-O. Goulet (ed.), *Le Commentaire entre tradition et innovation*, coll. 'Bibliothèque d'histoire de la philosophie' nouvelle série (Paris: Vrin, 2000), 411–24. On Godfrey of Fontaines, see Stephen D. Dumont, 'Godfrey of Fontaines and the Succession Theory of Forms at Paris in the Early Fourteenth Century', in Stephen Brown, Thomas Dewender and Theo Kobusch (eds), *Philosophical Debates at Paris in the Early Fourteenth Century* (Boston: Brill, 2009), 39–125.

5. Henrik Lagerlund, *Encyclopedia of Medieval Philosophy: Philosophy between 500 and 1500*, vol. 1 (New York: Springer, 2014), 553.

6. Cited in Pierre Duhem, *Medieval Cosmology: Theories of Infinity, Place, Time, Void, and the Plurality of Worlds* (Chicago: University of Chicago Press, 2011), 76.

7. Cited in Duhem, *Medieval Cosmology*, 76.

8. Lagerlund, *Encyclopedia of Medieval Philosophy*, 553.

9. Cited in Duhem, *Medieval Cosmology*, 85.

10. Carl Boyer, *The History of the Calculus and its Conceptual Development* (1949) (New York: Dover, 1959), 73.

11. Quoted in Marshall Clagett, *The Science of Mechanics in the Middle Ages* (Madison: University of Wisconsin Press, 1980), 214.

12. Sylla, 'Medieval Concepts of the Latitude of Forms', 263.

13. See Clagett, *The Science of Mechanics in the Middle Ages*, 335.

14. Clagett, *The Science of Mechanics in the Middle Ages*, 333.

15. Annelise Maier and Steven D. Sargent, *On the Threshold of Exact Science: Selected Writings of Anneliese Meier on Late Medieval Natural Philosophy* (Philadelphia: University of Pennsylvania Press, 2016), 39.

16. Nicole Oresme, *Tractatus de configurationibus qualitatum et motuum*. Translated as *Nicole Oresme and the Medieval Geometry of Qualities and Motions. A Treatise on the Uniformity and Difformity of Intensities Known as Tractatus de Configurationibus Qualitatum et Motuum*, ed. and trans. Marshall Clagett (Madison: University of Wisconsin Press, 1968), 167.

17. Oresme is clear that it does not matter which is which as long as the lines are perpendicular, but so as to be consistent he says he will 'let the extension of a quality be called its longitude and intensity its latitude or altitude'. Oresme, *Nicole Oresme and the Medieval Geometry*, 173 (Part 1, Chapter IV).

18. For a detailed account of Oresme and his difference with Descartes, see Gilles Châtelet, *Figuring Space: Philosophy, Mathematics and Physics* (Dordrecht: Springer, 2011).

19. Châtelet, *Figuring Space*, 42, my italics.
20. Just as Spinoza described in Book II of his *Ethics*. See Benedictus Spinoza, *A Spinoza Reader: The Ethics and Other Works*, trans. E. M. Curley (Princeton: Princeton University Press, 1994), Book II, Axiom 2 and Lemma 4, Lemma 7.
21. See Boyer, *History of the Calculus*; Sylla, 'Medieval Concepts of the Latitude of Forms'; and Duhem, *Medieval Cosmology*. Galileo gets these from the Oxford calculators. See Clagett, *The Science of Mechanics in the Middle Ages*, 346.
22. Max Jammer, *The Concepts of Force* (New York: Dover, 2012), 70–2.
23. Paul Murdin, *Full Meridian of Glory: Perilous Adventures in the Competition to Measure the Earth* (New York: Springer, 2008), 41.
24. See Galileo Galilei, *Dialogue Concerning the Two Chief World Systems – Ptolemaic and Copernican*, trans. Stillman Drake (Berkeley: University of California Press, 1962), 22–3 and 227, where the tunnel experiment is discussed. Also see Drake's 1974 translation of the *Discorsi* where Salviati presents 'experimental proof' of this postulate by pendulum motions. Stillman Drake (trans.), *Two New Sciences* (Madison: University of Wisconsin Press, 1974), 206–8, 162–4.
25. William Stahl (trans.), *Martianus Capella and the Seven Liberal Arts*, vol. 2, *The Marriage of Philology and Mercury* (New York: Columbia University Press, 1977), 332–3.
26. In mathematics, two quantities are in the golden ratio if their ratio is the same as the ratio of their sum to the larger of the two quantities. A Kepler triangle is a right triangle with edge lengths in geometric progression, following the golden ratio. Kepler defined this as a 'Divine Proportion', writing 'Geometry has two great treasures: one is the theorem of Pythagoras; the other, the division of a line into extreme and mean ratio. The first we may compare to a measure of gold; the second we may name a precious jewel.' As quoted in Mario Livio, *The Golden Ratio: The Story of Phi, the World's Most Astonishing Number* (New York: Broadway Books, 2002), 62.
27. 'I offer this work, the mathematical principles of philosophy, for the whole burden of philosophy seems to consist in this: from the phenomena of motions investigate the forces of nature, and then from these forces demonstrate the other phenomena.' Isaac Newton, Preface to the First Edition, *Sir Isaac Newton's Mathematical Principles of Natural Philosophy and His System of the World*, vol. 2, trans. Andrew Motte and Florian Cajori (Berkeley: University of California Press, 1962), xvii.
28. Newton, Definition III, in *Sir Isaac Newton's Mathematical Principles of Natural Philosophy*, vol. 2, 2.
29. Newton, Definition VIII, in *Sir Isaac Newton's Mathematical Principles of Natural Philosophy*, vol. 2, 5.
30. J. M. S. Pearce, 'Malpighi and the Discovery of Capillaries', *European Neurology*, 58.4 (2007), 253–5 (253).
31. Jole Shackelford, *William Harvey and the Mechanics of the Heart* (Oxford: Great Neck Publishing, 2006), 62.
32. Matteo Valleriani, *Galileo Engineer* (New York: Springer, 2013).

11

The Medieval Object II

Two more major scientific practices contributed to the development of intensive objects in the early modern world: the mathematics of infinity and the experimental method. This chapter argues that these new sciences followed a tensional pattern of motion.

Infinity

The third intensive science of the early modern world was the mathematics of infinity. Infinite objects are so strange and fascinating because one cannot entirely quantify them as with finite objects. Similar to kinematic and dynamic magnitudes, infinite objects are *intensive objects* because they are continually changing. One can always add one more number, indefinitely.

During the medieval and early modern periods, the mathematics of infinity became one of the most important sciences. In this chapter, I want to look closely at this new kind of object's kinetic conditions.

Actual infinity

Ancient and early modern scholars had different understandings of infinity. Aristotle argued that there were two kinds of infinity. What he called 'potential infinity' was the indefinite sequence of finite ordinal or cardinal objects. A potential infinity always has one more number that one has not yet counted. He called 'actual infinity' the totality or whole of an infinite sequence counted as 'one', as a cardinal number. Aristotle did not believe actual infinities were real.

For Plato and Aristotle, there was no such thing as a mathematics or science of processes. Science was the knowledge of discrete and static forms, not of their becoming or continuous qualitative deformation. Accordingly, they understood infinity as fundamentally divided between an ordinal form of unlimited sequence and a cardinal form of a finite whole. An object was *either* an unlimited ordinal *or* a finite cardinal. The idea that there could be an 'actual infinity' was like saying there could be an ordinal cardinal. It was impossible.

However, medieval and early modern mathematicians did not fully accept this division between two kinds of static objects. Instead, they developed a new method of quantifying a process-object. They defined this new object as a shifting coordination or ratio between ordinal and cardinal series. It was the mathematical equivalent of the physically dynamic and moving object. Some medieval and early modern mathematicians even borrowed the term 'actual infinity' from Aristotle, but gave it a new intensive definition. Below I want to show how this new understanding of 'actual infinity' as a real but *intensive* object emerged.

Infinite and infinitesimal

In the thirteenth century many scholars and natural theologians began thinking about nature's *movement* as an infinite and infinitesimal process. The third-century Hellenistic philosopher Plotinus had laid the groundwork for such an idea much earlier when he wrote that 'the One' was 'so great that its parts, too, are unbounded'. What if the discrete parts of things also had unlimited parts so small and numerous that they were impossible to count? God or nature would not be a simple cardinal 'one'. Instead, they would be ongoing coordinations of unlimited series, whose parts would also be endless series, and so on ad infinitum. This initial idea was enormously essential and utterly different than the cardinal notion of objects as wholes composed of parts. God is not a cardinal 'one', for Plotinus, but an infinite multiplicity of multiplicities.

The thirteenth-century Franciscan theologian and philosopher Richard of Middleton was one of the first to redefine Aristotle's idea of 'actual infinity explicitly' as a *mixture*. Middleton argued that actual infinity was not an infinity *in facto esse*, or 'actual fact', which is divided into discrete objects, but is instead an 'infinite *in fieri* (in act), which is what one calls the mixture of infinite in actuality with potentiality'.[1] In this radical new view, Middleton thought of actual infinity as the *immanent act or process* of counting and division itself.

In the same way that theologians defined God's *power* by his *act*, they started to think about God's infinity similarly. As the great French philosopher of science Pierre Duhem put it, 'The nature of the quantitative infinite is a mixture of actuality and potentiality.'[2] God's actual parts also contain actual parts and so on ad infinitum.

Middleton influenced the thirteenth-century Scottish philosopher and theologian Duns Scotus, who argued for a similarly new definition of actual infinity. 'What can exist in act is the measure of what can exist in

potentiality.'[3] Actual infinity, for Scotus, was in the *act* or process of counting. The ancients treated cardinal objects as fixed totalities of their parts,
but intensive objects, for Scotus, occur in the action of counting or holding
parts and wholes together.

After Scotus, the French philosopher Nicolas Bonet pushed the idea
of actual infinity *as action* to its extreme conclusion. Bonet reasoned that
if the power of the Prime Mover, or God, was infinite, then his power of
creation was unlimited in time or space. Therefore, each act of creating
a quality and a quantity could be 'produced simultaneously, particularly
since the production is the work of a power of infinite dimension'.[4] God
can coordinate many endless series and count them all simultaneously with
the infinite power of his action. For Bonet, this is the sense in which God's
power of counting is *actually infinite*.

Another fourteenth-century French scholar, Jean Buridan, developed
a beautiful geometrical illustration of this idea called the *linea gyrativa*, or
'spiral line'. The *linea gyrativa* was a cylinder whose height Buridan divided
into a sequence of proportional parts starting with 1/2, 1/4, 1/8, and proceeding into smaller and smaller divisions. Buridan intended these markings
to depict an extensive infinity of cardinal segments. Additionally, though,
he added a spiral line whose pitch was equal to the height of each segment
of the cylinder, as shown in figure 11.1. Buridan used this line to depict the
intensive infinity of qualitative change. The striking result of this shape was
that one could see how extensive and intensive series could be unlimited
and yet coordinated together in a single object.

In Italy, the great theologian Gregory of Rimini proposed a similar idea
in which God could create an actually infinite stone by adding a similarly
decreasing proportional mass (1/2, 1/4, 1/8 . . .) to a single stone. In an
hour, he would have an actually infinite stone, according to Rimini. From

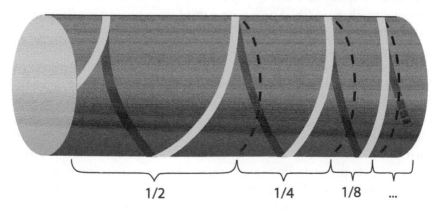

Figure 11.1 *Linea gyrativa.* Author's drawing.

Middleton to Gregory, the crucial new idea was that actual infinity was not a simple cardinal object but a continual act of coordination between unlimited series. Both series become actually infinite through the process of counting and coordination.

If this sounds similar to the latitude of forms, there is a good reason for that. The English cleric and scholar Thomas Bradwardine explicitly connected the existence of actual infinity with 'infinitesimal' changes in latitude. In his theory, the line of latitude of change in a quantity was not discrete atoms but infinitely small continuous dimensions (*Nullum continuum ex indivisibilibus infinitis integrari vel componi*).[5]

The great theorist of latitude Nicole Oresme also used this idea to describe the continuous rate of change of a moving object. He divided the distance the object travelled into intervals and imagined an object that slowed by half its speed over each interval. To determine the average speed, he summed the following series: $1/2 + 3/8 + 1/4 + 3/16$, and so on ad infinitum.[6] As the series approached zero, it became infinitesimal.

Over the fifteenth and sixteenth centuries, the concept of the infinitesimal spread across Europe.[7] The German philosopher and theologian Nicholas of Cusa used it to invent an 'infinite geometry' in which he defined a circle as a polygon with infinite and infinitesimal sides. He described human knowledge of the divine as an asymptotical *terminus ad quem* to be approached by the act of proceeding to the limit.

The genuinely transformative idea in all of this was the invention of a new kind of object made by coordinating two infinite series. What made the series infinite was not the objects, but the *act* or *power* of holding them together in tension. God's *power* to count is not reducible to the objects he creates and coordinates. In the language of the trinity, the intensive object was a triple, trefoil or triune object that coordinated an unlimited series of qualitative changes (the Father) with an unlimited series of quantitative changes (the Son) using a third dynamic series of action (the Holy Spirit).[8]

There is another beautiful image of the kinetic *tension* holding the infinite object together in the work of the Italian mathematician Bonaventura Cavalieri. Cavalieri imagined infinitesimals as threads that make up an infinite cloth.[9] Just as one makes a cloth by a coordinated tension between vertical and horizontal threads, actually infinite objects were made by the coordination of different unlimited series. The Italian physicist and mathematician Evangelista Torricelli developed a similar idea for an 'infinite geometry'. He defined a line as an infinity of infinitesimal points, a plane as a series of infinitesimal lines, and solids as a series of infinitesimal planes.[10] In such a theory, one is no longer dealing with finite ordinal or cardinal objects but instead with ongoing tensions between indefinite series.

Calculus

One of the most significant developments in the mathematics of infinity was calculus. Calculus emerged from the ideas of the latitude of forms and actual infinity. The twin inventors of calculus were the German philosopher Gottfried Wilhelm Leibniz and the English scientist Isaac Newton. They built on past work and made the tremendous synthetic leap to bring latitude and infinity together in a single analytic and coordinate-based system. Calculus allowed mathematicians to perform arithmetical and algebraic operations on geometrically infinite objects with variable rates of change or curvature.

Such operations were impossible in Descartes' coordinate system because finite arithmetical ratios were only possible with rectilinear lines. As Descartes fully admitted:

> Geometry should not include lines (or curves) that are like strings, in that they are sometimes straight and sometimes curved, since the ratios between straight and curved lines are not known, and I believe cannot be discovered by human minds, and therefore no conclusion based upon such ratios can be accepted as rigorous and exact.[11]

This is the problem calculus aimed to solve.

Fluxus

Leibniz and Newton took the idea of mapping a geometrical curve as a continuous path of a moving object from kinematics. As such, one could coordinate an object at each point along the x and y metric axes of Descartes' coordinate system. One could quantify a curved line by continuously coordinating its horizontal and vertical axes along a path. Just as Oresme determined the rate of change in motion as an infinite sum of all the changes over time, calculus determined the area below the curving line, or 'integral', through an infinite sum of rectangles of infinitesimal width. Calculus also determined the rate of variation or slope of the curve by an infinite sum of ratios over time. Leibniz called this the 'differential'.

By treating curves like moving quantities, or *fluens*, one could find the instantaneous rate of change, or *fluxion*, of the curve. The Oxford calculator Richard Swineshead was the first to propose the idea of treating curves as trajectories. It was then popularised by Torricelli and adopted by the English mathematician Isaac Barrow. Barrow then passed the idea on to his student Isaac Newton, who adopted it in his calculus in *Methods of Fluxions* (1671; published 1736). Newton wrote that his calculus's aim was, 'given an equation that consists of any number of flowing quantities, to find the fluxions: and vice versa'.[12] This approach's strength was that it no longer limited one to a static and rectilinear model of geometrical analysis, such as in the work of Descartes, Pierre de Fermat or John Wallis. With Newton's model, one could find the integral and rate of change in a quantity over time for curved lines.

However, what was especially interesting from a kinetic perspective was that the act of doing such calculations as they approached infinity showed the role that mathematical activity itself played in the 'final' solution. If one can always approximate a curve closer and closer, where do you stop? Newton's calculus dealt explicitly with 'flowing quantities' and posited a continuum of infinitesimally smaller rates of change within each rate of change.

Therefore, the idea of an actual infinity appeared in Newton's work insofar as each finite part contained an infinity that one could sum to infinity. Flowing quantities, for Newton, were infinities made of infinities coordinated with other infinities.

Function

Leibniz and Newton both used the analytic geometrical system axes to coordinate an infinite and infinitesimal series. Leibniz called this coordination between the infinite and the infinitesimal a 'function'. In a 1673 letter to the Swiss mathematician Johann Bernoulli, Leibniz defined a function as a quantity related to a curve, such as a curve's slope at a specific point. A function is the coordination of input along one axis to output along another. Functional coordination follows a constant rule or 'tension'.

Oresme had done the same thing by coordinating variations in speed with variations in time to discover a constant or variable 'rate of change'. The rate of change in velocity or a slope is a function. It has a fixed proportional rate at which x varies alongside a proportional or tensionally linked ratio with y. A mathematical function is a linkage or continuous coordination of one series to another. Subject, function and series are three aspects of the same act.

Infinitesimals

Newton and Leibniz also incorporated infinitesimals into their early calculus methods as a way to deal with the problem of a continuous rate of change in the slope of a curve. This was because a geometric curve is not reducible to a finite series of rectilinear segments. One way to understand a curve is to imagine it as composed of an infinity of infinitesimal, non-zero regions. There is an infinity of in-between slopes between any two parts of a slope that can be measured.

For example, in figure 11.2, determining the tangent line at P allows us to see that as the slope increases along the x-axis towards the other secant point Q, the tangent line also increases along the y-axis in a certain

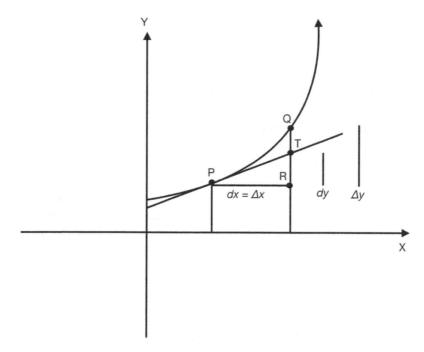

Figure 11.2 Tangent to the curve at P. Author's drawing.

proportional ratio. Leibniz formulated this tensional ratio as dy/dx, or 'the continuous change (Δ) in y with respect to x'. Leibniz called this the 'differential'.

Every geometrical line, straight or curved, can also be metrically or extensively divided infinitely. Therefore, to determine the curve's slope between the secant PQ, we require an infinite sum of infinitesimal intervals.

Early on, Newton and Leibniz both used unique symbols to indicate this infinitesimally small but non–zero interval. Newton used a slanted zero (o), and Leibniz used a 'delta' or 'd', meaning 'change', which we use today. Eventually, Newton and Leibniz both conceded that such non–finite infinitesimal quantities were, as Leibniz wrote, merely 'useful fictions'. Leibniz wrote that he 'did not believe at all that there are magnitudes truly infinite or truly infinitesimal'.[13] The link between these fictions and reality 'he undoubtedly felt would be found in his law of continuity, which he had taken as basic in all his later work in the calculus', as the historian of mathematics Carl Boyer writes.[14]

In his *Principia*, Newton also later wrote that

ultimate ratios in which quantities vanish, are not, strictly speaking ratios of ultimate quantities, but limits to which the ratios of these quantities decreasing without limit, approach, and which, though they can come nearer than any given difference whatever, they can neither pass over nor attain before the quantities have diminished indefinitely.[15]

From Plotinus to Newton, I hope I have shown that what was 'actually infinite' for these thinkers was not an ordinal or cardinal object or series of objects. In that sense, they agreed with Aristotle that an actual *cardinal* infinity would be a contradiction. Instead, actual infinity was an ongoing mixture and coordination between two indefinite series. This is an intensive object, and its coordination is what I called 'many-to-many' in Chapter 9. For these thinkers, there were no static ordinals or cardinals. All objects were in motion and intensive. Objects were not discrete but rather coordinated ratios, tensions, functions, proportions and rates of infinitesimal and infinite series.

Experimentation

The fourth principal intensive science of the early modern world was experimental methodology. Experimental method was the science of scientific practice. It was a set of steps or stages for how to proceed towards knowledge. What was significant about this was that it took the constitutive *act* of making and reproducing scientific knowledge itself seriously. The *action* of the scientist became part of the *object*.

During the middle ages, the rise of artisan craftwork, instrument design and mechanics led to a divergence from Aristotelian syllogistic or deductive logic. Craftspeople and earlier engineers did not tend to start with axiomatic truths and deduce the postulates. Instead, they began with practical problems and used trial and error, or induction, to develop best practices. The craftsmen and scientists who created new instruments for measuring intensive objects were more interested in the ancient materialist philosophers such as Democritus, Epicurus and Lucretius, who believed that sensation was the foundation of knowledge. During the early modern period, materialist philosophy made a comeback in natural philosophies where people took matter and motion seriously.[16]

The Latin words for 'experimentation' were *experientia* and *experimentum*, from the Latin word *experior*, 'to experience, to do or act', plus the suffix *-mentum*, meaning 'something resulting through an instrument or medium'. The scientific method combined three aspects of knowledge: the subject of sensuous experience, the method or medium of the experience and the result of the experience. Scientific knowledge was co-created by all three. The Latin word *instrumentum*, from *īnstruō*, 'to build, construct; arrange', also emphasised the active and constitutive nature of scientific investigation through a medium. There is no purely passive or objective place where scientists can view objects. Their actions and environment are intimately entangled aspects of objects and knowledge.

The objects of experimental science were 'tensional' because they triangulated a triple tension between subject, instrument and object. Francis Bacon and others did not base the scientific method on Aristotelian mental deduction but an active *medium of knowledge*. Knowledge was not something one merely discovered, but something one co-created. The early moderns reinterpreted Aristotle's deductive logic or *organon*, the Greek word for 'instrument', as more than a passive lens to see the world, but something constructed and revisable like any other instrument. Scientists did not deduce knowledge but instead produced and experienced it as a parallax between subject, apparatus and object.

Experimentation, like other intensive objects, was in motion. The experimenter had to establish and continually maintain an experimental set-up using instruments affected by the experiment and by the experimenter. Setting up or reproducing an experiment involves stabilising and recording a particular relationship between subject, object and instrument. To study changing objects in motion, one's instruments and subjects also need to move. Experimentation is the continuous coordination of all three movements.

The proliferation of scientific instruments such as the compass, telescope, microscope, pendulum and clock created an increasing kinetic tension between the subjects and objects of knowledge. In response to their understanding of how instruments mediated subjects and objects, scientists aimed to manipulate this medium with a new method.[17]

Let us first examine the early rise of mechanical experimentation and how these new techniques became formalised in the scientific method.

Mechanical experimentation

The scientific methodologies developed by Islamic and Latin scholars in the middle ages had their origins in the rise of *practical mechanics* and the proliferation of *technical instruments*. During the middle ages, technicians modified ancient implements such as the astrolabe, sundial, map and compass rose, and medieval instruments such as the telescope, thermometer, microscope, pendulum and mechanical clock, to coordinate their intensive movements with an extensive or metric dimension. Using these instruments, scientists could record moving objects and patterns along the two axes.

Scholars did not deduce these instruments from first principles. They were developed through trial and error by mechanics, artisans and miners to solve practical problems. Over time, these instruments were adopted and adapted by the canonical 'men of science' into tools of knowledge.

The experimental method began with these mechanics and artisans, who developed a three-part procedure. They designed their tools in a particular environment, tested them there through experience, then adjusted their instruments and repeated the process. The outcome of this informal method was that mechanics could produce repeatable and accurate ratios of coordination between the instrument's qualitative changes and the quantitate markings on it. When appropriately calibrated, the two axes of quality and quantity could record the movement of intensive objects. As one axis changed, the other should vary in direct or inverse proportion. Calibrating instruments involves a series of mutual adjustments between subject, object and tool.

Francis Bacon, Galileo Galilei, Robert Boyle and Isaac Newton all built their work on a long history of instrument manufacture and calibration going back to the ancient and early medieval world. In particular, Galileo, Newton, the Dutch physicist Christiaan Huygens and the English polymath Robert Hooke all claimed that they built their own instruments. However, the basis for their instruments was not original. For the most part, they made improvements to existing ones.

These scientists also learned their skills from the artisans of their time. It was the partnership between the 'greatest scientists' and the 'best of the craftsmen', as the historian David Landes writes, 'that finally gave us precise time-keepers to the measure of the stars and the sea; and toward the end, when the scientists thought they had done as much as they could, it was the craftsmen who persisted and completed the task'. Artisans thus 'possessed surprising theoretical knowledge and conceptual power'.[18]

The famous French philosopher of science Alexander Koyré claimed that the scientific method was the product of 'pure unadulterated thought',[19] but this is incorrect. Artisans did not play a passive material role, while scholars played an active and formal role as inventors. Contrary to the idealist position that 'the genesis of the scientific revolution is in the mind',[20] it was manual labourers and mechanics who developed an informal experimental method based on induction, refinement and calibration. They were the first to start coordinating the qualitative and quantitative aspects of instrumental measurement.[21] As the historian Clifford Conner argues, Bacon's *Novum Organum* (1620), 'representing the culmination of this tradition, did not introduce a new approach to gaining scientific knowledge; he was describing a social phenomenon that had begun centuries earlier and was advocating that it be systematically exploited'.[22]

Galileo, Bacon and the natural philosopher William Gilbert all explicitly claimed that their inspirations came from miners, sailors, blacksmiths, foundrymen, mechanics, lens-grinders, glassblowers, clockmakers and shipwrights of their era.[23] In astronomy, Tycho Brahe and Johannes Kepler relied on the method of triangulation that was invented earlier by the Dutch surveyor Gemma Frisius.[24] Galileo says in *Dialogues Concerning Two New Sciences* (1638) that the 'constant activity' at Venice's weapons factory, the 'famous arsenal',

> suggests to the studious mind a large field for investigation, especially that part of the work which involves mechanics; for in this department all types of instruments and machines are constantly being constructed by many artisans, among whom there must be some who, partly by inherited experience and partly by their own observations, have become highly expert and clever in explanation.[25]

The influence of mechanics on science was explicit in Francis Bacon's admiration of the 'mechanical arts'. For instance, he wrote that the university-based sciences 'stand like statues, worshipped and celebrated, but not moved or advanced', while 'the mechanical arts . . . having in them some breath of life, are continually growing'.[26] In contrast to the Swiss physician Paracelsus, however, Bacon was much more interested in harnessing and formalising this knowledge in the service of a renewed scholarly apparatus than in the mechanics' intellectual advancement and workers themselves as leaders in epistemological innovation.

The scientific method

Diverse methods of calibration, design and mechanics abounded in the middle ages. Bacon did not invent them, but he did synthesise and formalise them in his book, *Novum Organum Scientiarum* (*The New Instrument of Science*) (1620). I want to look at three core aspects of his experimental method and show how each follows the tensional pattern of motion common to all intensional objects. Those three features are induction, instrumentation and coordination.

Induction

The first and perhaps best-known aspect of Bacon's experimental method was his rejection of Aristotelian deductive logic and his affirmation of a new logic called 'eliminative induction'. Aristotle's logic began with a priori axioms about the true nature of things and then centrifugally unfolded them into an array of diverse consequences. However, Bacon argued in favour of a new kind of induction that 'elicits axioms from sense and particulars, rising in a gradual and unbroken ascent to arrive at last at the most general axioms; this is the true way, but it has not been tried'.[27]

Both deduction and induction 'start from sense and particulars, and come to rest in the most general; but they are vastly different . . . one forms certain abstract and useless generalities from the beginning, the other rises step by step to what is truly better known by nature'.[28] Furthermore, if these general axioms were found to be contrary to experience 'it would be more truthful to amend the axiom'.[29]

Against the centrifugal and 'circular' operation of deduction, whose propositions and axioms merely confirm one another without generating any strictly 'new' knowledge, Bacon proposed a more tensional or 'triangular' technique of instrumental induction instead. In this new induction, the subject, instrument and object all varied continually in a proportional, diffracted and linked relationship with one another. Bacon's induction was not merely the linear production of axioms from experience. It was a continuous transformation of a whole process, including subjects, instruments and objects, as they vary over time and in tension with one another.

For Bacon, an experimental truth was different from a deductive truth. Induction was not an inversion of deduction. It was a new kind of object. Instead of beginning from the central aristocratic polis and radiating truth outwards upon the senses and masses, Baconian induction began with heterogeneous peripheries of experience in the countryside, the laboratory, the university, the mines and the seas. From their divergent locations and experiences, Bacon hoped to identify common truths that held these unique experiences together and distinguished them. As people's concrete experiences changed, Bacon imagined that new common truths would emerge, as long as we listened to them.

This was how knowledge production and revision worked in the mechanical arts for centuries. Sensation came first. Then one built an instrument or apparatus. Afterwards, someone revised the device through qualitative and quantitative coordination, which then revealed new patterns, which then called for new or altered tools. Bacon's idea for 'eliminative induction' was that truths would emerge from the tensions between divergent techniques. People would eliminate what was false and did not work.

Instrument

The second aspect of Bacon's experimental method was his theory of the instrument. This was one of his most significant contributions to the new intensive objects of early modern science. From the very first aphorisms of Book I of *Novum Organum*, Bacon was explicit that 'the new instruments of science' were not like the old deductive instruments of the mind, but were material kinetic objects that provided the medium (*-mentum*) through which knowledge was produced (*īnstruō*).

> Neither the bare hand nor the unaided intellect has much power; the work is done by tools [*instrumentis*] and assistance, and the intellect needs them as much as the hand. As the hand's tools [*instrumenta*] either prompt [*cient*] or guide its motions, so the mind's tools [*instrumenta*] either prompt or warn the intellect.[30]

Just as the movement of the hand needs instruments to put it into motion or action (*cient*), so the mind needs tools to carry (*suggerunt*) the intellect (*mentis intellectui*) forward. In this way, Bacon used the 'mechanical arts' as the model for the intellectual arts. For Bacon, instruments served a double role in science as practical and intellectual objects. A subject without an instrument was not worth much (*multum valet*) because the intensive object was one that appeared *as a process*. The subject was simply a process of intensive and extensive coordination with the instrument and object.

The intensive object, Bacon said, appeared only as a 'flux and alteration of matter' (*fluxu materiae*).[31] Therefore, it could only be approached by experiment and indirectly through a medium in tension with an instrument. For Bacon, the instrument does not give us unmediated access to objective nature. Our senses touch only the experiment to some degree (*sensus de experimento tantum*), which in turn only touches or experiences nature to some degree (*experimentum de natura*), thus actively shaping or determining (*judicat*) the thing itself. The subject touches the experiment, and the experiment touches nature. Together, the three are held together and apart in a tensional structure. As one changes, the whole *experimentum* shifts.[32]

Coordination

Now we reach the heart of Bacon's method. The third aspect of Bacon's experimental method was the coordination between intensive and extensive series. In Book II of his *Novum Organum*, Bacon proposed a specific method for determining truths from sensuous experience and experimentation. The method consisted in creating three tables, each containing an indefinite and ongoing history of instances in which a phenomenon, such as heat, always occurs ('existence and presence', Table 1), never occurs ('absence', Table 2), or occurs to some degree (Table 3). 'So tables must be drawn up and a coordination of instances [*tabulae et coordinationes instantiarum*] made, in such a way and with such organisation that the mind may be able to act upon them.'[33]

Bacon called each of these tables an 'ordered' (*ordine*) 'experimental history' (*Historia vero naturalis et experimentalis*). Like the phenomena themselves, these tables could go on indefinitely.[34] The purpose of producing an ordered history like this was to discern the relations and common structure of a phenomenon more easily. Bacon accomplished this by creating a horizontal line across the three tables that coordinated the object's changes in appearance.

Table 1 functioned like an extensive series of discrete 'present' instances, Table 2 acted like the gaps or non-instances that separated these discrete presents, and Table 3 functioned like the intensive or continuous degrees of the phenomenon's appearance. At the core of Bacon's method was the coordinated tension between extensive and intensive series or 'histories'. The object of scientific knowledge was neither reducible to a finite ordinal or cardinal object nor an ordinal or cardinal series. Instead, it occurred as an intensive process–object of coordinated or intersecting objects.

Bacon's method of coordination also included the instrument's mediation and the revision of the inductive process. The process of knowledge production continuously mutated in tensional proportion with each of these aspects. Intensive objects are not eternal forms but rather structural laws of ratio and coordination between material changes. As Bacon writes in Aphorism LI in Book I of *Novum Organum*,

> The human understanding is carried away to abstractions by its own nature, and pretends that things which are in flux are unchanging. But it is better to dissect nature than to abstract; as the school of Democritus did, which penetrated more deeply into nature than the others. We should study matter, and its structure [*schematismus*], and structural change [*meta-schematismus*], and pure act, and the law of act or motion; for forms are figments of the human mind, unless one chooses to give the name of forms to these laws of act.[35]

Bacon did for the mind what mechanics did for bodies and machines. He set up a calibration system or structural coordination between extensive and intensive aspects of objects. This was the genius of his scientific method.

Conclusion

This chapter concludes the third section of this book. My argument in this section was not that all the sciences of the middle ages and early modern period were identical or accomplished the same thing, but that each followed in its way a tensional pattern of motion. However, around the middle of the eighteenth century, this new intensive science was increasingly overtaken by another kind of object I call the 'potential' object. The description of this new object and its historical sciences is the subject of the fourth and final section of Part II of this book.

Notes

1. Cited in Pierre Duhem, *Medieval Cosmology: Theories of Infinity, Place, Time, Void, and the Plurality of Worlds* (Chicago: University of Chicago Press, 2011), 47.
2. Cited in Duhem, *Medieval Cosmology*, 81, n. 22.
3. Duns Scotus, *Sentences*, Book 3, Dist. 13.
4. Cited in Duhem, *Medieval Cosmology*, 108, n. 83.
5. Cited in Carl Boyer, *The History of the Calculus and its Conceptual Development* (1949) (New York: Dover, 1959), 67.
6. Boyer, *History of the Calculus*, 86.
7. See Boyer, *History of the Calculus*; Duhem, *Medieval Cosmology*; Paolo Zellini, *A Brief History of Infinity* (London: Penguin, 2005).
8. See Thomas Nail, *Being and Motion* (Oxford: Oxford University Press, 2018), ch. 27.
9. Bonaventura Cavalieri, *Exercitationes Geometricae Sex* (1647) (Bononiæ: Typis Iacobi Montij, repr. 1986), 239–40.
10. Boyer, *History of the Calculus*, 124–9.
11. René Descartes, *Geometry of René Descartes* (New York: Dover, 2012), 91.
12. Brian Clegg, *A Brief History of Infinity: The Quest to Think the Unthinkable* (London: Constable, 2003).
13. Gottfried W. Leibniz, *Opera Philosophica* (Aalen: Scientia Verlag, 1974), vol. III, 500.
14. Boyer, *History of the Calculus*, 219.
15. Isaac Newton, *Sir Isaac Newton's Mathematical Principles of Natural Philosophy, and His System of the World: Volume One: The Motion of Bodies*, trans. Andrew Motte, ed. Florian Cajori (Berkeley: University of California Press, 1962), 39.
16. In fact the rise of early modern science and the experimental method was influenced by the atomists, among other factors. See Pierre Vesperini, *Lucrèce: Archéologie d'un classique européen* (Paris: Fayard, 2017).
17. However, towards the end of the seventeenth century and into the eighteenth, scientists began using the term *observationes* or 'observations', to describe situations where nature was more passively seen instead of actively experimented with. According to the German natural philosopher Gottfried Wilhelm Leibniz, 'there are certain experiments that would be better called observations, in which one considers rather than produces the work'. Cited in Lorraine Daston and Elizabeth Lunbeck, *Histories of Scientific Observation* (Chicago: University of Chicago Press, 2011), 86.
18. Cited in Clifford Conner, *A People's History of Science: Miners, Midwives, and Low Mechanicks* (New York: Nation Books, 2009), 260.
19. Cited in Conner, *A People's History of Science*, 270.
20. Cited in Conner, *A People's History of Science*, 102.
21. For full development of this thesis, see Edgar Zilsel, Robert S. Cohen, Wolfgang Krohn, Diederick Raven and Joseph Needham, *The Social Origins of Modern Science* (Dordrecht: Kluwer, 2003); see also Pamela Smith, *The Body of the Artisan: Art and Experience in the Scientific Revolution* (Chicago: University

of Chicago Press, 2012), 239: 'Laid the foundations for a new epistemology, a new *scientia* based on nature.'

22. Conner, *A People's History of Science*, 283.
23. Conner, *A People's History of Science*, 276.
24. Conner, *A People's History of Science*, 257.
25. Cited in Conner, *A People's History of Science*, 285.
26. Conner, *A People's History of Science*, 250.
27. Francis Bacon, *The New Organon* (Cambridge: Cambridge University Press, 2000), Book I, Aphorism XIX, 36.
28. Bacon, *The New Organon*, Book I, Aphorism XXII, 37.
29. Bacon, *The New Organon*, Book I, Aphorism XXV, 37.
30. Bacon, *The New Organon*, Book I, Aphorism II, 33.
31. Bacon, *The New Organon*, Book I, Aphorism XVI, 35–6.
32. As Grosseteste suggested with his geometrical and optical theory of knowledge. See Thomas Nail, *Theory of the Image* (Oxford: Oxford University Press, 2019), 192–3.
33. Bacon, *The New Organon*, Book II, Aphorism X, 109.
34. Bacon, *The New Organon*, Book II, Aphorism X, 109.
35. Bacon, *The New Organon*, Book I, Aphorism LI, 219.

IV THE POTENTIAL OBJECT

12

The Elastic Object

The most unique and powerful kind of object in the modern world was the 'potential' object. What I call 'potential objects' were the most abstract of all because one cannot reduce them to any single state or series. Instead, modern scientists treated them as *possible* objects with a *range* of yet-to-be-determined states.

These objects exist in a strange hypothetical future where some are more likely to appear than others. Tracking their history allowed scientists to study and anticipate the range of what they might do in the future. The potential object is like a falling die whose possible states all coexist in mid-air. The moment the die hits the ground, and one side faces up, it disappears. The potential object becomes actual.

A potential object is a chance object in the dual sense of the Latin word *cadentia*, meaning 'to fall or happen'. It is something thrown into motion which produces a singular occurrence. The potential range of possibilities is real because it based on the actual occurrences, but it is not actual in the same way as the occurrences.

Although the science of potential objects began to blossom around the middle of the eighteenth century, it had its origins in something much earlier: dice. Dice are fascinating objects because of their dramatically dual nature as both *potential* and *actual* objects. They are potential in that all their sides coexist as possible outcomes and actual in that only one can face up at a time. Dice were the first potential objects because they expressed a yet-to-be-determined amount that people could determine through the movement of throwing.

The oldest dice come from ancient Mesopotamia, around 3000 BCE,[1] and people played dice games throughout ancient Egypt, Greece and Rome.[2] However, it was not until the Italian polymath Girolamo Cardano and the French mathematicians Blaise Pascal and Pierre de Fermat started doing formal studies of dice outcomes in the sixteenth and seventeenth centuries that it became a science.

Before Cardano, people treated dice as sources of unpredictable divination. Aristotle, for example, distinguished between determinate causes and indeterminable or spontaneous causes. The first, he said, could be actual objects of knowledge, but the second could not:

> That which is per se cause of the effect is determinate, but the incidental cause is indeterminable, for the possible attributes of an individual are innumerable . . . It is necessary, no doubt, that the causes of what comes to pass by chance be indefinite; and that is why chance [Τύχη, *tyche*] is supposed to belong to the class of the indefinite and to be inscrutable to man.[3]

The idea that indeterminate objects were inscrutable to scientific knowledge persisted until around the middle of the seventeenth century.[4] In the ancient world, the only knowable objects were actual, discrete, ordinal or cardinal objects. In the middle ages and early modern periods, the movements of objects could be known but only as determinate proportions of change. Intensive objects still could not deal with unpredictable or possible objects such as dice. The modern object was a different kind of object because it was contingent and indeterminate. Many modern sciences even treated the x- and y-axes themselves as indeterminate. Potential objects were an *indeterminate range of quantities* that one could study before their actualisation as ordinal, cardinal or intensive objects.

This new kind of object emerged around the middle of the eighteenth century and spread through all the twentieth century's sciences. In the next two chapters, I want to show how this object emerged with the rise of statistical sciences, digital logic and transcendental mathematics. First, though, in this chapter, I want to provide a synthetic definition of the modern object's core kinetic features.

Elastic motion

What is elastic motion, and how does it describe the movement of potential objects? Elasticity is the ability of an object to expand and contract across a range of possible states. How much can we transform an object before it breaks or becomes something else? How much can we change a whole field of related objects before it falls apart and the relations change? These are the kinds of questions posed by modern science.

It is a strange way to think about things we normally consider relatively static and discrete. The intensive sciences were so radical because they could record the qualitative and quantitative changes of objects over time.

However, in doing so, they had assumed that the coordinate plane itself was static and discrete. What if the x- and y-axes were not straight lines, but curved? One of the most brilliant innovations of modern science was to treat the geometrical plane of space and time itself as an elastic and deformable field, like the surface of a balloon.

Modern science saw that neither objects nor their planes of coordination were stable. Objects were ranges and distributions that coexisted in what scientists called 'superpositions', 'sheaves', 'groups' or 'power sets'. These new objects contained the full range of permutations of a potential object. They were parameters of how far particular objects could be stretched or changed before becoming something different or breaking. Each actual object was treated as only one chance occurrence or contracted state of a wider elastic object.

How big was the field of possible permutations? It, too, was elastic. The set of all possible occurrences of an object can expand or contract. For example, in set theory, the 'power set' of a set S $\{x, y\}$ is $\mathscr{P}(S) = \{\{\}, \{x\},$ $\{y\}, \{x, y\}\}$. The power set shows each subset of the set, including the 'empty set', or the subset without any elements. This mathematical field is elastic because one can add or subtract objects from it without adding another field. The same field can accommodate numerous changes.

Transforming fields is a kinetic and elastic action where humans modulate how expanded or contracted the field of objects is. Fields are mutable, and the range of objects can be arbitrarily large or small, as the intensive sciences discovered with infinite objects. There is no universal answer to what one should include in a set and what one should not. It is a real act of knowledge that one must continually perform. The total possible range of an object cannot be known in advance but requires practical experimentation. How big is a 'big enough' sample-size to make predictions? It all depends.

As new objects or permutations occur, one can keep adding them to the elastic set of possible objects indefinitely. One never knows when a completely unexpected event might happen. For example, people used to think that the set of all swans was white. Let w stand for white swans and S $\{w\}$ be the set of all swans, thus $\mathscr{P}(S) = \{\{\}, \{w\}\}$. However, with the discovery of black swans in Australia, the set of all possible swans was elastically stretched to include black swans without ceasing to be 'the set of all possible swans': S $\{w, b\}$ and $\mathscr{P}(S) = \{\{\}, \{w\}, \{b\}, \{w, b\}\}$. In this way, potential objects were practically open and elastic.

Transcendental quanta

The next defining feature of potential objects is what I call their uniquely 'transcendental quanta'. By 'transcendental', I mean what the German philosopher Immanuel Kant called the 'conditions of possibility' for something. By 'quanta', I mean, following the Latin word *quantus*, the 'how muchness' of something. 'Transcendental quanta' is an appropriately strange term for an object of 'possible conditions for quantity'. One of the amazing powers of modern science was that it focused its attention on objects' environmental parameters. What is the nature of the background upon which objects exist, move and change? The potential sciences studied these conditions of possible objects as objects in their own right.

Euclid assumed that his world of geometrical objects was homogeneous and flat. Under such conditions, parallel lines would never meet. However, if space were heterogeneous and curved, it would change the conditions for possible quantities. It would not only mean that another geometry was possible but that many new geometries were possible. Non-Euclidean geometries, elastica theories, Minkowski spaces and general relativity introduced new 'hyperobjects' into the sciences by treating space and time themselves as elastic objects.

Potential objects are superpositions of a range of possible objects simultaneously. Actual numbers or quantities are not merely selections from these ranges but elastic contractions of them. In an event, the potential becomes actual. For example, the set of all possible swans expands and contracts as actual swans are counted and observed. However, no matter how big or small the set is, it will always be 'the set of all possible swans'.

Modern science also treated potential objects as sets of *discrete* objects. This is because they extrapolated the potential object like a collage from the actual occurrences of these objects in the world. They observed the world, counted each occurrence of an object, and then imagined a potential object that was the simultaneous existence of all possible discrete occurrences. The sciences of potential objects assumed that, since actual objects occur as *discrete events*, their conditions are superpositions of discrete possibilities.

Kant did the same. Although he intended to describe universal transcendental structures for all humanity, his model was based on particular Eurocentric and historical assumptions.[5] In this way, transcendental objects were always retroactive constructions based on the practical limitations of how many occurrences one observes over some finite time. Since the lengths and places of observations change, potential objects are always expanding and contracting their ranges. Potential objects are always provisional.

The insight of the potential sciences was that objects had more primary processes over a range of space and time. However, their limitation was that they restricted these processes to discrete occurrences and on occasion postulated a total set of such occurrences.

Many-to-one coordination

Potential objects also have a unique structure of coordination. Imagine again the modern object *par excellence*: the die. Imagine that we never see the die, but only the series of outcomes of the throw. Over a certain amount of time, the numbers 1 to 6 appear many times. Each throw of the die is unique, but we can concretely count it as a discrete number. Since we never see the die, it is always possible that a 7 might appear, but practically, since we only see numbers 1–6, we accept those as the possible range of numbers.

What I am calling a 'many-to-one' coordination is when we take a long series of occurrences and imagine them as coexisting 'sides' or 'dimensions' of a single potential object. This is easy to visualise with a six-sided die, but now imagine that all objects are like dice with numerous dimensions. Now imagine the entire universe as one big die with many sides. This is the largest potential object. No one has ever seen the whole universe. Still, it is possible to extrapolate from the starting assumption that if objects are discrete occurrences, then the universe is the 'power set' of all possible objects. The idea of objects as discrete occurrences and potential objects as totalities of all possible occurrences go hand in hand.

This modern technique was built on an inversion of the ancient method of one-to-many coordination. The classical Greeks first assumed a cardinal unity's abstract existence and then deduced from this axiom a range of self-evident truths about the world. In classical Greece, many thinkers saw particular objects as imperfect copies of a perfect original model. By contrast, potential objects were not perfect models of simple unity, nor were they intensive or dynamic objects. They were composites of coexisting fragments without convergence, like a Picasso painting. Potential objects were what mathematicians called 'group-objects', where all permutations, reflections and rotations of an object coexist.

In this way, the potential object is not a single static object but an assemblage of heterogeneous occurrences coordinated to a single multidimensional hyperobject. No amount of particular occurrences will ever exhaust the full range of what a potential object can do. Scientific knowledge of a potential object is, therefore, always limited in nature. Yet as long as one assumes that objects are discrete, self-identical and self-enclosed occurrences, then there must be a single total potential object of all these discrete pieces. This is the philosophical assumption at the heart of modern science.

Albert Einstein famously wrote that '[God] does not play dice with the universe', but later clarified this statement by saying that 'God tirelessly plays dice under laws which he has himself prescribed.'[6] For Einstein, the world changes but the laws do not. It was no coincidence that Einstein framed his cosmology in terms of *dice throws*. It is a perfect image of modern science. For Einstein, the universe was a finite and closed system with a complete set of laws for all possible discrete occurrences. Reality was the actualisation of certain superimposed possibilities.

After many discrete occurrences have been recorded and coordinated into a single potential object, modern science can then reverse the logic again. One can then treat the potential object as if it were its whole history of various occurrences. These 'many-objects' then 'fall' and become one concrete object at a time. The object begins with a wide range of possibilities but then contracts into a single occurrence. One can then record each event and add it back to the expanded potential object.

The potential object is like a hydrological cycle. Many water surfaces on the earth evaporate into a single sky, which then rains down many discrete drops of water, which then evaporate back into the sky. The same potential multiplicity expands and contracts, composes and decomposes, but without losing anything of what it is.

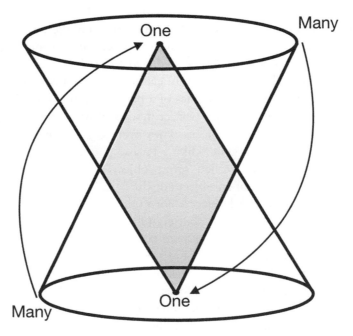

Figure 12.1 Double cones of probability. Author's drawing.

Conclusion

Potential objects are strange because they are not single objects or even a series of single objects but ranges of superimposed possible objects. Their ranges expand and contract over time as new occurrences happen, but the totality of discrete objects does not change. The scientific knowledge of such potential objects is not absolute but practical and experimental. As long as time moves on, we never have the whole set of objects. We have to gather the occurrences of objects and expand the range of possible states without knowing when the range is complete. In this sense, the potential object is fundamentally temporal.

This chapter has tried to provide a synthetic outline of the potential object's three main features. In the next two chapters, I want to look more closely at modern science's material history to see how these patterns emerged concretely.

Notes

1. David Schwartz, *Roll the Bones: The History of Gambling* (Las Vegas: Winchester Books, 2013).
2. Florence David, *Games, Gods and Gambling: A History of Probability and Statistical Ideas* (Mineola, NY: Dover, 1998).
3. Aristotle, *Physics*, trans. R. P. Hardie and R. K. Gaye (New York: Dover, 2017), Book II, ch. 5.
4. Before this time, for example, the Latin word *probabilis* simply described a plausible or reasonable opinion or action. R. C. Jeffrey, *Probability and the Art of Judgment* (Cambridge: Cambridge University Press, 1992), 54–5. See also J. Franklin, *The Science of Conjecture: Evidence and Probability Before Pascal* (Baltimore: Johns Hopkins University Press, 2015), 113, 126.
5. For a longer discussion of Kant's philosophy and its basis in history and movement, see Thomas Nail, *Being and Motion* (Oxford: Oxford University Press, 2019).
6. Albert Einstein, autograph letter signed 'A. Einstein' to Paul Epstein, Princeton, n.d., but before November 1945. See <https://www.livescience.com/65697-einstein-letters-quantum-physics.html> (last accessed 4 March 2021).

13

The Modern Object I

Modern objects follow a predominantly elastic pattern of motion. In the next two chapters, I want to show how this shared pattern of movement emerged historically among several modern sciences: the statistical sciences, digital logic and transcendental mathematics. These sciences were not always in direct or indirect conversation with one another. Nonetheless, there was a fascinating historical resonance where a shared pattern of motion and understanding took hold. I want to begin the story of this resonance by looking at several sciences that all adopted the framework of statistical mechanics.

Statistical mechanics

Statistical mechanics is the organisation of objects as ranges of *probable distributions*. There were two key historical discoveries at the heart of this idea. The first was that more powerful instruments allowed scientists to see or posit much larger populations of particles than had previously been imagined, such as molecules, atoms, genes and isotopes. The second was that the movement of these large populations was seemingly random. It was unfeasible to apply Newton's equations to molecules one by one. Early modern scientists could track intensive objects because they were fewer and macroscopic. However, to study large populations of objects and their group movements, scientists introduced a new method.

Potential objects emerged from this pursuit to describe and predict the movement of large numbers of apparently randomly moving bodies. Objects that scientists had before treated as solids they increasingly treated as swarms of randomly moving microscopic bodies. Solidity was an effect of a more elastic expansion and contraction of swarming bodies in an equilibrium state. Instead of mapping every tiny moving body, statistical mechanics let scientists track tendencies or probabilities of population movement.

This chapter looks at how four major statistical sciences followed this elastic pattern of motion: chemistry, electromagnetism, thermodynamics and particle physics.

Chemistry

Chemistry was one of the earliest modern sciences. Before the mid-eighteenth century, scientists treated gases and fluids as tensional substances composed of relatively static objects linked together in ether. For Torricelli, Boyle and Newton, there was only a handful of different but tensionally homogeneous kinds of elements locked in specific ratios with one another. However, throughout the modern period, this idea gave way to a new kinetic theory of gases and an elastic theory of 'conservation of mass'.

The kinetic theory of gases

In 1738 the Swiss mathematician Daniel Bernoulli published *Hydrodynamica*, which laid the basis for the kinetic theory of gases. Bernoulli argued that gases and other 'elastic fluids' were composed not of static molecules locked in tension, but of massive, uncountable populations of tiny 'corpuscles' moving immeasurably fast and in random directions. For Bernoulli, pressure was the impact of these corporeal motions exerted on a vessel, and heat was the energy these bodies released through movement. Bernoulli wrote,

> Let the cavity contain very minute corpuscles, which are driven hither and thither with a very rapid motion; so that these corpuscles, when they strike against the piston and sustain it by their repeated impacts, form an elastic fluid which will expand of itself if the weight is removed or diminished . . .[1]

What is kinetically crucial here is that Bernoulli defined objects as elastic and 'continuous'[2] because they were composed of an indeterminate quantity of *discrete* parts, whose random motions expanded and contracted in equilibrium. At the sunrise of modern science, Bernoulli's defined elasticity by discreteness. This had a significant effect on other theories of elasticity, as I show in this chapter. Elasticity was a net kinetic effect created by the expansion and contraction of enormous populations of discrete objects.

The conservation of mass

In 1789 the French chemist Antoine Lavoisier, known as the 'Father of Chemistry', further advanced the modern attack against the older chemistry based on the four classical elements. Against the existence of phlogiston, the substance of fire, Lavoisier argued that there were many discrete and indivisible physical 'elements' that composed objects. He wrote that 'If we apply the term *elements*, or *principles of bodies*, to express our idea of the last

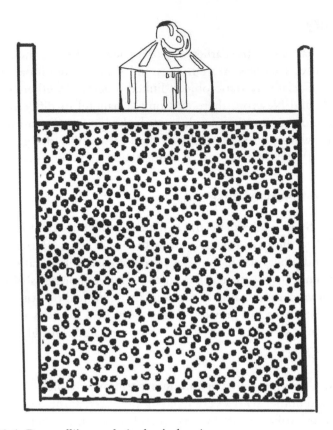

Figure 13.1 Bernoulli's vessel. Author's drawing.

point which analysis is capable of reaching, we must admit, as elements, all the substances into which we are capable, by any means, to reduce bodies by decomposition.'[3] For Lavoisier and nineteenth-century chemists such as the Russian Dmitri Mendeleev, an element was the smallest decomposable object that composed all other objects.

By carefully measuring each chemical's mass and confining their interactions in a sealed vessel, Lavoisier was able to see that the total mass of chemicals after a reaction did not change. Elements were neither created nor destroyed in chemical processes but merely elastically expanded, contracted, composed and decomposed – just as Bernoulli had initially suggested in his kinetic theory. This was the physical basis of the law of conservation of mass.

In 1802 the English chemist John Dalton, following Bernoulli and Lavoisier's idea of chemical compounds' discrete elasticity, wrote a paper called 'On the Expansion of Elastic Fluids'. In this paper, he argued that the elasticity of elements in a compound fluid occurred in whole-number ratios because they were made of discrete atoms. For Dalton, matter occurred in proportional relations because these discrete elastic objects expanded and contracted in specific proportions. Although the movement of these objects was ultimately stochastic, they remained *statistically proportional* in a closed system at equilibrium. In this way, chemistry became a science of elastic fluids of discrete particles.

Electromagnetism

The next science of elastic objects was electromagnetism. This was an extraordinary science. For the first time, scientists treated electricity and magnetism as elastic fluids with *potential* kinetic or *actual* energy. In contrast to the middle ages' tensional theories which understood magnetism and electricity as rigidly linked coordinations between determinate objects across empty space, the modern theory of electromagnetism treated them as *elastic fluids*.

Electromagnetic fluids

The Danish physicist Hans Christian Ørsted experimentally confirmed the direct relationship between electricity and magnetism in 1820. However, it was the English scientist Michael Faraday and the Scottish scientist James Clerk Maxwell who added a whole new elastic interpretation of electromagnetism as 'lines of force' or 'wave propagations' through a universal elastic medium of *aether*. In contrast to thinking of electricity and magnetism as two separate fluids, Faraday and Maxwell thought of them as two aspects of the same kinetic object 'transmitted by vibrations through an . . . infinitely elastic ether'.

For Faraday, 'the infinite elasticity assumed as belonging to the particles of the aether' provided a universal medium for gravity, light and electromagnetism. 'The lines of force', therefore, 'represent the assumed high elasticity of the aether.' The 'condition of the aether', for Faraday, was explicitly like 'the state of the rare gas' in chemistry.[4] Therefore, electromagnetism was defined by the same principles of equilibrium, conservation and elasticity that Lavoisier had used to describe chemistry.

Maxwell often seemed to share the idea that elasticity assumed discrete objects. He wrote, 'It is often asserted that the mere fact that a medium is elastic or compressible is a proof that the medium is not continuous, but is composed of separate parts having void spaces between them.'[5] Although Maxwell ultimately decided to leave the question of the continuum or discreteness of the aether 'to the metaphysicians',[6] he still insisted on the existence of the aether as a requisite elastic medium through which 'electric elasticity'[7] was possible. 'The medium [aether], in virtue of the very same elasticity by which it is able to transmit the undulations of light, is also able to act as a spring.' This idea 'enables us to resolve several kinds of action at a distance into actions between contiguous parts of a continuous substance [aether]'.[8] Electromagnetism, therefore, shared a similarly elastic pattern of motion to that of chemistry.

Electromagnetic potential

Electromagnetism was the study of potential objects. It was initially Bernoulli who introduced the idea of a difference between 'potential' and 'actual' energy in 1738.[9] Following the Dutch physicist Christiaan Huygens' distinction between a pendulum's energy states, Bernoulli was the first to introduce the idea of 'potential' energy into chemistry. Potential energy is the quantity of possible energy that an object can exert if released or set into motion, such as in the fall of a pendulum. Just as when one releases a pendulum weight, and its potential energy is converted into kinetic energy as it approaches its lowest energy point at equilibrium, chemical compounds can also release potential energy through reactions.

Bernoulli's idea of a 'quantum of *possible* energy' was taken up by other fields, including electromagnetism, by Joseph Louis Lagrange in 1773, Adrien-Marie Legendre between 1784 and 1794, Pierre-Simon Laplace from 1782 to 1799, Siméon Denis Poisson in 1813, George Green in 1828, Carl Friedrich Gauss in 1840 and, most importantly, by James Clerk Maxwell in 1864.[10]

An object with a variable amount of potential energy and an actual amount of kinetic energy is a distinctly modern kind of object. For Aristotle, an object's potential had its end built into its essence, as an acorn becomes an oak. However, the potential of modern objects depended on their *environmental conditions*. The higher the pendulum was before it dropped, or the greater an electrical current, the greater their potential energies. Furthermore, as an object expresses its potential in the world as actual kinetic energy, it changes the environment. As the environment changes, so does the potential energy. It was a feedback loop, like the positive and negative sides of a magnetic field. The potential object expands and contracts around points of equilibrium.

For example, an electromagnetic potential is the energy needed to cross a threshold of expansion or contraction, attraction or repulsion. The electromagnetic potential is a kind of 'transcendental' object because it is the condition for the possibility of the expansion or contraction of an elastic fluid. As an electromagnetic object changes, so does its potential or threshold for new kinds of relations. The English scientist Sir William Thomson

(Lord Kelvin) predicted this reciprocal reflex action between potential and actual states in 1853: 'The phenomena require us to admit the existence of a principal discharge in one direction, and then several reflex actions backward and forward, each more feeble than the preceding, until the equilibrium is obtained.'[11]

This theory of reflex action was later confirmed and elaborated by Maxwell, whose 'results seem to show that light and magnetism are affections of the same substance, and that light is an electromagnetic disturbance propagated through the field according to electromagnetic laws'.[12] 'Light', Maxwell argued, 'consists in the transverse undulations of the same medium which is the cause of electric and magnetic phenomena.'[13] 'But the medium, in virtue of the very same elasticity by which it is able to transmit the undulations of light, is also able to act as a spring.'[14] Maxwell developed equations that unified and made explicit what had been implicit in the theory of potential in dynamics, chemistry, electricity, magnetism and light. The relation between potential and actual energy was an *elastic* relation defined by oscillations around an equilibrium point.

Electromagnetic objects were not just actual objects but potential fields of flux that shaped objects and relations. The electromagnetic field was like a fabric or fluid in which bodies emerged and floated like corks. Treating objects as 'fields' was a brilliant and powerful idea that influenced general relativity and quantum field theory.

Thermodynamics

The science of heat and energy followed the same kinetic pattern of elasticity. The combination of Bernoulli's kinetic theory of gases and Lavoisier's elemental theory made it possible for scientists to think of all physical matter as a collection of randomly moving objects following the elastic law of conservation. The movement of heat was more akin to a fluid field than any single stable object. However, the fundamental problem was how to measure the motion and energy of this uncountable number of unobservable bodies in elastic liquids and solids. This was the daring challenge of thermodynamics.

The kinetic theory of matter

The first significant work on thermodynamics was completed by the German physicist Rudolf Clausius in 1857. Clausius revived Bernoulli's kinetic theory of gases and extended it to solids and liquids in his essay, 'On the Nature of the Motion which We Call Heat'. He did this by proposing that tiny particles of heat had oscillatory and rotational motions in addition to the linear movements that Bernoulli described. Clausius also extended his theory to the 'molecular motions occurring in solids, liquids, and gaseous states'.[15] All three kinds of objects could be measured by the 'mean velocity . . . for the totality of molecules'[16] in a single object. This was a brilliant move. Instead of calculating the motion of objects one at a time as in Newtonian mechanics, Clausius calculated the statistical movements of entire populations of swarming objects.

Clausius's work directly influenced Maxwell, and Maxwell saw the potential in it. He added to Clausius's statistical method a unique mathematical formalism that replaced the idea of a merely 'average' rate of total speeds of objects with a *statistical distribution* of varying molecular velocities. This new equation treated the molecules bouncing around in an object as having a range of velocities. For Maxwell, potential energy did not just have an *average speed* but a range of distinct possible speeds. The collective movement of a population of molecules colliding with one another at varying speeds eventually resulted in an equalisation of temperatures and created a tendency towards equilibrium motion. Maxwell's equations described this tendency beautifully.

Maxwell also described them as explicitly elastic kinds of motion that were continually expanding and contracting around an equilibrium point. Objects, for Maxwell, were statistical ranges of possible discrete states in random motion. The same process occurs in gaseous, fluid and solid bodies. Solid bodies are just like fluids with less elasticity to their form. 'In solid bodies', Maxwell wrote, 'the elasticity of form appears in many cases to be smaller in proportion to that of volume.'[17] Maxwell spelled out clearly an early version of the kinetic theory of objects developed later by Einstein. Maxwell wrote that

> The dynamical theory supposes that the molecules of solid bodies oscillate about their positions of equilibrium, but do not travel from one position to another in the body. In fluids the molecules are supposed to be constantly moving into new relative positions, so that the same molecule may travel from one part of the fluid to any other part.[18]

Entropy

In 1868 the Austrian physicist Ludwig Boltzmann added another crucial consideration to Maxwell's theory of statistical distribution. Boltzmann saw that Maxwell had left out the energetic effects that the bouncing particles had on one another as they collided. This was a brilliant insight but was a challenging process to model mathematically. It required a knowledge of all particles' statistical velocity distribution and the frequency and force of their collisions over time. Without ever having seen an atom, many scientists at the time felt this was purely speculative. However, Boltzmann's genius was that he modelled the average velocity of the total particle collisions *statistically* instead of measuring the actual concrete collisions of invisible atoms. He then added this to Maxwell's statistical equation for the force, mass and time of particle motions and drew several important conclusions.

In 1872 Boltzmann formulated the results in his famous 'H-theorem'. He showed that the total set of increasingly disordered collisions in a closed system was the mechanism that returned a system to an equilibrium. As their randomly moving particles collided, objects tended to become increasingly disordered over time due to their motion. As formulated by Boltzmann, entropy was a strong statistical tendency of all closed systems over time.

Einstein's great contribution to the kinetic theory of matter brought together Boltzmann's statistical model with the conservation of mass. In a famous 1905 paper, Einstein showed that the diffusion of an object undergoing random or 'Brownian' motion diffused at a particular rate called the 'mean squared displacement'. By correlating the total weight of a gas with the rate of the diffusion of another substance in that gas, Einstein was able to determine the size and number of particles in the fluid. Einstein proved that the particles that Clausius, Maxwell and Boltzmann had imagined were real. For the first time, he showed a way to experimentally measure them based on their statistical movements as a random and elastic population. This new method worked for gases, liquids and solids, and confirmed the elastic nature of objects.

Physics

Modern physics followed in tandem the same statistical and elastic pattern. Perhaps most radically, physics expanded atomic and classical field theories into a fully relativistic field theory of space and time.

Particles

Building on others' work, Einstein showed beyond dispute that a potential object was something composed of discrete particles whose random motions expanded and contracted towards equilibrium. What were these fundamental particles, though, and did they have their own internal motions? In 1897 the British physicist J. J. Thomson discovered that cathode rays travelled through air much faster than atoms. This led him to postulate the existence of 'corpuscles' 1,000 times smaller than an atom. He later called these 'electrons'. To this discovery, the British–New Zealand physicist Ernest Rutherford added another in 1911 when he was shooting radiation through a piece of gold foil. As he observed the deflection angles off the surface, he and his collaborators noticed that the deflection angle suggested that the radiation was bouncing off something small and heavy inside the atoms of gold. He called this the 'nucleus' of the atom.

Thomson's and Rutherford's experiments painted a new picture of the atom as a tiny solar system. The atom's mass was concentrated in a heavy central nucleus surrounded by orbiting electrons like planets around a sun. In this way, physicists understood atomic objects as more elastic, since they expanded and contracted as they gained and lost electrons.

In 1900 the German physicist Max Planck made a similar discovery about electromagnetic fields. Faraday and Maxwell had treated electromagnetic fields as elastic fluids composed of 'particles of the aether' in which larger, 'ponderable' particles floated. However, Planck was able to make a precise measurement of the possible discrete states of the potential field. Planck formalised what Faraday and Maxwell had only conjectured.

Planck found that he could compute the total energy of the electromagnetic waves radiating from a hot body by treating it as discrete 'quanta', or small, brick-like packets. He found that the size of the energy packets depended on the frequency of the electromagnetic waves.[19] This allowed him to describe fields as elastic surfaces composed of quanta of energy, smaller than atoms. This was the first formula of quantum theory.[20]

Five years later, in 1905, Einstein showed that light was also composed of discrete quanta, called 'photons'. Following Planck, Einstein showed that when specific frequencies of light hit atoms, the light caused the electrons orbiting the atom to break loose and generate electrical energy. Einstein called this the 'photoelectric effect'. What caused the electrons to break loose was not the total number of photons that hit the electrons but the precise *frequency* or *quanta of the energy packets*, which Einstein calculated using Planck's constant. Einstein's paper on the photoelectric effect was published in the same year as his paper on the kinetic theory of matter, and is part of the same modern theory of potential objects. From chemistry to quantum physics, modern science identified smaller and smaller discrete objects whose collective statistical expansions and contractions moved like elastic surfaces of potential energies.

In 1911 the Danish physicist Niels Bohr brought together Planck's and Einstein's method of quantisation. Bohr wanted to know why certain atoms reflected the specific spectrums of light they did. Why are objects the colour that they are? Physicists at the time knew that colour-frequencies related to the electron orbits around the nucleus of the atom. Still, no one knew why the electrons did not collapse into the nucleus or produced the colours or discrete orbital patterns that they did.

Bohr's insight was to apply the method of quantisation to the energy levels of these orbiting electrons. In doing so, he discovered that electrons only orbited at certain discrete distances from the nucleus.[21] Bohr also saw that electrons do not move continuously between their orbits around the atom's nucleus but instead seem to disappear entirely and 'leap' from one orbit to another. Furthermore, each orbiting electron's discrete energy frequency determined the frequency or colour of light that reflected off them. Atoms, he concluded, reflected different colours of light depending on the orbits of their electrons. What was also intriguing about Bohr's discovery was that electrons' position and energy were not static but fluctuated across a range of possible states. In this way, Bohr's work deepened the statistical and potential nature of modern objects. All that seemed solid melted into potential ranges of fluctuating discrete objects.

What was the relationship between this *potential* range of energies and their *actually* observed electron orbits? In 1925 the German theoretical physicist, and Bohr's colleague at the Copenhagen Institute, Werner Heisenberg argued that the relationship was fundamentally *uncertain*. One can only measure a subatomic particle's movement at a discrete position in space and instant in time when an observer shines a light to see it. However, since light is made of photons, the photons collide with the electron and alter its course. The more an observer wants to determine the electron's precise position, the more light they shine on it, and the more it appears as a discrete body at a particular position in space surrounded by photons.[22] Conversely, the less the observer interacts with the subatomic particle, the less discrete the particle's position appears, and the more the object appears as a potential range of possible momentum. Therefore, Heisenberg concluded, because observation always affects what one is observing, one cannot determine the position and momentum of a subatomic particle simultaneously.

Heisenberg's 'uncertainty principle' perfectly expresses the structure of what I am calling the potential object. Knowledge of such an object occurs along a spectrum between possible and actual. Modern scientists treat the possible as a superposition of discrete states in a closed system, and the actual like a snapshot of one possible state.

Particle-wave duality

At the heart of modern physics is the idea of a 'complementarity' between the object as a statistical range or 'wave' of possible paths and a determinate particle. Physicists call this 'particle–wave duality'. Modern physicists developed several fascinating 'slit experiments' designed to demonstrate this duality.

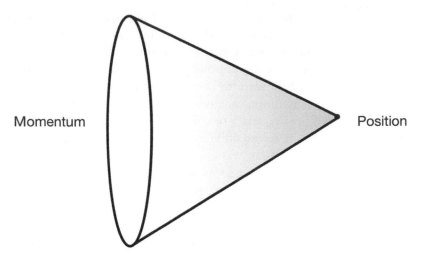

Momentum   Position

Figure 13.2 Quantum uncertainty cone. Author's drawing.

For example, what would happen if one fired photons at a wall through a single-slit opening? In classical physics, we would expect that the particles would move uniformly through the slit and create an outline of the opening on the wall on the other side. However, when the British physicist Thomas Young did this experiment in 1801, he found that the photons spread out after they went through the slit and created a scatter pattern on the wall, as if a wave had passed through the slit. When Young shone a light through two slits, it produced a diffraction pattern of alternating areas of concentration, as if a single wave had gone through both slits simultaneously and created two overlapping wave patterns on the wall.

However, if Einstein was correct that light was made of discrete photons, how could one go through both slits simultaneously? They cannot, and further experiments confirmed that each photon only went through one slit. But then why does this still produce wave patterns?

For most modern scientists, the phrase 'particle-wave duality' or 'complementarity' indicates the strange way that one could describe light as *both* a particle *and* a wave.[23] As the American quantum physicist Richard Feynman was fond of saying, 'the double-slit experiment has in it the heart of quantum mechanics. In reality, it contains the only mystery.'[24]

Particle-wave duality is a beautiful image of the modern object. It is both *expanded* across a range of possible positions or 'wave patterns', and elastically *contracted* in a single position or 'particle'. Possibility and actuality are merely two sides of the same potential object.

This same elastic theory of the object continued into later twentieth-century physics, where physicists treated particles simultaneously as

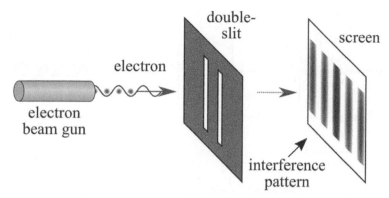

Figure 13.3 Double slit experiment. https://en.wikipedia.org/wiki/Double-slit_experiment#/media/File:Double-slit.svg.

distributed 'wave functions'. What we now call quantum 'wave mechanics' first emerged in the work of the French physicist Louis de Broglie. De Broglie argued that if light worked like a particle-wave, then electrons might also. In his 1924 dissertation, de Broglie used Planck's constant to show that the discrete quantum of energy in an electron was proportional to the wave-energy of its frequency.

Although de Broglie interpreted this wave function as a deterministic description of how different particles must necessarily move, Max Born and Einstein both reinterpreted it as a *statistical* theory. They found that an electron's energy and frequency shaped the statistical likelihood that an actual electron was more *likely* to be observed in a given orbit. Just as Bohr showed that the frequency of light waves determined colour, so Born and Einstein showed that a potential electron *wave's frequency and energy statistically shaped an electron's position.*

However, de Broglie's model of the atom, just like Maxwell's model of gases, did not consider the interaction of covariant factors. So just as Boltzmann added this to Maxwell's gas equations, so the Austrian physicist Erwin Schrödinger did the same for de Broglie's wave equations. In 1926 Schrödinger interpreted position and momentum not as a standing probability wave but as a dynamic wave that moved and changed in time. This new wave function could be combined with other wave functions to show their interaction and shifting ranges of particles.

The mathematics Schrödinger proposed was not strictly probabilistic but rather deterministic partial differential equations with multiple unknown variables. Schrödinger had taken de Broglie's deterministic waves and superimposed them on one another in a closed system to represent the total set of deterministic momentums and positions before their interactions.

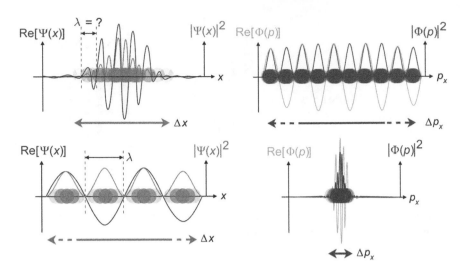

Figure 13.4 De Broglie wave probabilities. Position x and momentum p wave functions corresponding to quantum particles. The colour opacity of the particles corresponds to the probability density of finding the particle with position x or momentum component p. Top: If wavelength λ is unknown, so are momentum p, wave-vector k and energy E (de Broglie relations). As the particle is more localised in position space, Δx is smaller than for Δpx. Bottom: If λ is known, so are p, k and E. As the particle is more localised in momentum space, Δp is smaller than for Δx. https://en.wikipedia.org/wiki/Wave–particle_duality#/media/File:Quantum_mechanics_travelling_wavefunctions_wavelength.svg.

However, in 1928 the English physicist Paul Dirac developed an equation that made all the variables in the wave covariant and interacting. This included not just position and momentum but velocity, energy and angular momentum. Instead of assuming a superposition of deterministic states, as Schrödinger did, Dirac treated free-particle interactions as a series of probability amplitudes for calculating whether this or that value of a variable would appear in the next interaction.[25]

This is Dirac's quantum mechanics: a recipe for calculating the spectra of the variables, and a recipe for calculating the probability that one or another value in the spectrum appears during an interaction. That's it. What happens between one interaction and the next is not mentioned in the theory. It does not exist. The probability of finding an electron or any other particle at one point or another can be imagined as a diffuse cloud, denser where the probability of seeing the particle is stronger.[26]

For Dirac, possibility and actuality alternate in a series of interacting transitions, unlike Schrödinger's frozen 'wave'.

However, potential waves are not like the actual classical waves you find at the beach. They are probability ranges of possible outcomes. Possible states are individually actualised like throws of a die when one observes a particle. An expanded range of many potential states contracts or 'collapses' into a single particle.

These physicists did not all agree with one another's interpretations of quantum mechanics. However, they all shared a belief in the elasticity of the potential object. Potential objects are statistical objects with complementary potential and actual states. At one end, modern physics treats objects as expanded 'probability clouds' of simultaneously possible states. At the other end, objects are contracted in discrete particle-objects.

The so-called 'deep mystery' of quantum mechanics arose directly from understanding the real existence of a genuinely elastic object that is *both* possible *and* actual *at the same time*. It is a decidedly modern problem. If we assume that objects are fundamentally discrete 'quanta', we can imagine an equally real total set of possible discrete states. Then the object becomes dual.

Relativity

Two of the most radical new objects of modern science were space and time themselves. In his 1905 'special theory' of relativity, Einstein argued that space and time were not, as Newton thought, 'absolute' or static aspects of reality. They were two aspects of the same elastic field. According to Einstein, space and time were *relative* or elastically related to each other. There is no single absolute unit of duration outside space or any single unit of space outside time. In his 'special theory of relativity', Einstein argued that the faster a body moves, the slower time would go relative to everything else. The passage of time slows down or speeds up depending on how fast something is moving through space.

Einstein contrasted his theory with Newton's mechanical conception of space as a series of mechanically linked and immobile bodies of what Einstein termed 'quasi-rigid' ether.[27] Instead, Einstein claimed that his 'space-time theory and the kinematics of the special theory of relativity were modeled on the Maxwell–Lorentz theory of the electromagnetic field'.[28] In other words, Einstein wrote, 'the whole change in the conception of the ether which the special theory of relativity brought about, consisted in taking away from the ether its last mechanical quality, namely, its immobility'.[29] There is no absolute, static or preferred frame of reference from which to measure the space and time of any moving body in relativity.

In Einstein's 1915 'general theory of relativity', all moving bodies, including Newton's, Faraday's and Maxwell's aether bodies, move through an even more foundational elastic field of spacetime,[30] with no 'rigid reference-bodies' at all.[31] Spacetime, Einstein argued, was a 'surface of constant curvature', bending, stretching and undulating like the curves of a giant flexible 'mollusk'.[32] Just as Maxwell's fields of elastic electromagnetic fluid produced waves, ripples or wavelengths, so Einstein's gravitational field produced 'gravitational waves'. The elasticity of spacetime was impossible in Newton's tensional mechanics because Newton assumed that bodies interacted at an infinite speed.

Kinetically speaking, Einstein's theories of relativity offered a transcendental theory of spacetime as a universal and elastic background field. The expansions and contractions of objects are all relative to one another precisely because spacetime itself does not have a fixed Euclidean structure but curves and stretches continuously like a sheet of gravitational rubber.

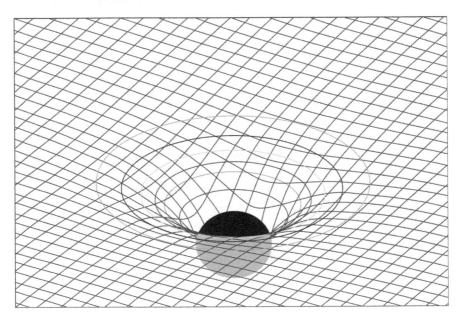

Figure 13.5 Curved spacetime. https://en.wikipedia.org/wiki/Maxwell%27s_equations_in_curved_spacetime#/media/File:Gravitation_space_source.svg.

Conclusion

In this chapter, I have shown that modern statistical mechanics invented the study of potential objects, which followed an elastic pattern of motion. In the next chapter, I want to look at two more modern sciences that similarly contributed to developing this new kind of object: digital logic and transcendental mathematics.

Notes

1. Daniel Bernulou, 'The Kinetic Theory of Gases', in James R. Newman (ed.), *The World of Mathematics* (New York: Dover, 2000), 774.
2. Bernulou, 'The Kinetic Theory of Gases', 774.
3. Antoine-Laurent Lavoisier, *Elements of Chemistry in a new systematic order, containing all the modern discoveries*, trans. Robert Kerr (New York: Dover, 1789), xxiv.
4. Michael Faraday, 'Thoughts on Ray-Vibrations', *Philosophical Magazine*, S.3, 28.188 (1846), 345–50, available at <http://www.padrak.com/ine/FARADAY1.html> (last accessed 4 March 2021).
5. James Clerk Maxwell, *The Scientific Papers of James Clerk Maxwell*, vol. 2 (Mineola, NY: Dover, 2003), 774.
6. Maxwell, *The Scientific Papers*, vol. 2, 323.
7. Maxwell, *The Scientific Papers*, vol. 2, 141.
8. Maxwell, *The Scientific Papers*, vol. 2, 323.
9. Daniel Bernoulli and Johann Bernoulli, *Hydrodynamics, or Commentaries on Forces and Motions of Fluids* (1738) (New York: Dover, 1968), 11, 30. See also Olivier Darrigol, *Worlds of Flow: A History of Hydrodynamics from the Bernoullis to Prandtl* (Oxford: Oxford University Press, 2009), 4–5.
10. Irina Markina, 'Potential Theory: The Origin and Applications', available at <http://folk.uib.no/ima083/courses_files/potential.pdf> (last accessed 4 March 2021).
11. *Proceedings of the American Philosophical Society*, Vol. II, No. 23, August, Sept. & Oct., 1842 (Philadelphia: Printed for the Society by John C. Clark, 1842), 193.
12. James Clerk Maxwell, 'A Dynamical Theory of the Electromagnetic Field', *Philosophical Transactions of the Royal Society of London*, 155 (1865), 459–512.
13. J. J. O'Connor and E. F. Robertson, 'James Clerk Maxwell', *School of Mathematics and Statistics, University of St Andrews*, archived from the original on 28 January 2011, available at <http://www-groups.dcs.st-and.ac.uk/~history/Biographies/Maxwell.html> (last accessed 4 March 2021).
14. Maxwell, *The Scientific Papers*, vol. 2, 323.
15. Rudolf Clausius, 'On the Nature of the Motion which We Call Heat' (1857), reprinted in Stephen Brush, *Kinetic Theory: The Nature of Gases and of Heat, Volume 1* (New York: Elsevier, 2013), 111.
16. Clausius, 'On the Nature of the Motion which We Call Heat', 131.
17. James Clerk Maxwell, 'On the Dynamical Theory of Gases', in Maxwell, *The Scientific Papers*, vol. 2, 26–78 (27).
18. Maxwell, 'On the Dynamical Theory of Gases', 27.
19. For waves of frequency v, every quantum, or energy packet, has energy ($E = hv$), where h is a new fixed numerical quantity, the quantum of electromagnetic action, which relates the energy carried by a photon to its frequency.

20. What became and still is called 'the old quantum theory' after quantum mechanics was introduced in 1925.

21. The discrete increments were measured in Planck's constant, h.

22. Heisenberg ultimately adopts a theory of quantum interaction, and abandons the idea of epistemological relativism often associated with his uncertainty principle. 'Following a heated discussion wherein Bohr offers an important criticism of Heisenberg's analysis, Heisenberg acquiesces to Bohr's point of view. Though it is little discussed, Heisenberg includes an admission of these important shortcomings of his analysis in a postscript to his famous uncertainty paper. In an important sense, this postscript constitutes an undoing of the analysis that he presents in the body of the text, and yet this erroneous analysis has become the standard exposition on the reciprocity relations. The uncertainty principle continues to be taught to students and spoken of by physicists and nonphysicists in accord with Heisenberg's account when by his own admission his account had been based on a fundamental error. Ironically, there is no mention of Bohr's account of the reciprocity relations, that is, the indeterminacy principle. Indeed, if Bohr's contributions to these discussions are mentioned at all, it is usually with a historically respectful nod to complementarity; but even this is seldom mentioned anymore.' Karen Barad, *Meeting the Universe Halfway: Quantum Physics and the Entanglement of Matter and Meaning* (Durham, NC: Duke University Press, 2007), 301.

23. There are 'both/and' and 'neither/nor' interpretations of the double slit experiment. By far the most common is the 'both/and' interpretation. However, in Part III I argue instead for an 'indeterminate' interpretation, following the American physicist Karen Barad.

24. Richard Feynman, *Feynman Lectures on Physics, Vol. I: The New Millennium Edition: Mainly Mechanics, Radiation, and Heat*, ed. Robert B. Leighton and Matthew Sands (New York: Perseus Hachette, 2011), ch. 37 on 'quantum behavior'.

25. See Carlo Rovelli, *Reality Is Not What It Seems: The Elementary Structure of Things*, trans. Simon Carnell and Erica Segre (New York: Penguin, 2017), 120–6.

26. Rovelli, *Reality Is Not What It Seems*, 124.

27. Albert Einstein, *Aëther und Relativitätstheorie* (Berlin: Verlag von J. Springer, 1920). English trans. 'Ether and the Theory of Relativity', available at <http://www-history.mcs.st-andrews.ac.uk/Extras/Einstein_ether.html> (last accessed 4 March 2021). See also L. Kostro, *Einstein and the Ether* (Montreal: Apeiron, 2000); and J. Stachel, 'Why Einstein Reinvented the Ether', *Physics World*, 14.6 (2001), 55–6.

28. Einstein, *Aëther und Relativitätstheorie*.

29. Einstein, *Aëther und Relativitätstheorie*.

30. 'But unlike Newton's space, which is flat and fixed, the gravitational field, by virtue of being a field, is something that moves and undulates, subject to equations – like Maxwell's field, like Faraday's lines.' Rovelli, *Reality Is Not What It Seems*, 81.

31. Albert Einstein, *Relativity: The Special and General Theory*, ed. Robert W. Lawson, Robert Geroch and David C. Cassidy (New York: Plume, 2006), 26.

32. Einstein, *Relativity*, 125–8.

14

The Modern Object II

This chapter looks closely at two more significant sciences that shaped the modern object's elastic structure: digital logic and transcendental mathematics. This chapter argues that these new sciences shared a common theory of 'potential objects' despite their differences.

Digital logic

Digital logic was a modern science of the binary digits, 1 and 0. By coding objects with a series of binary digits, scientists and engineers could create machines that followed simple instructions written in this language. For example, one could translate the number '9' into the binary code '01001' or a lower-case 'a' could be translated into '1100001'. The more binary digits in the series, the wider the range of possible objects that one could communicate. For instance, binary strings of eight digits can represent 256 unique possible objects.

Digital objects are 'potential' objects because they are total sets of discrete objects where each actual object is only one in the total range of possible objects. In this way, digital logic is a closed communication system of discrete possible objects.

The first technology to begin using digital logic was the punch card system, invented in 1801. A punch card was a piece of paper with a range of places where one could punch a hole. Based on the pattern of holes punched in the card, machines such as typewriters and mechanical looms could perform specific actions without direct human operation. Engineers could reduce many differences to a single difference between 1 and 0. The nineteenth-century English polymath Charles Babbage was the first to imagine that these punch cards could be used as operating instructions to perform complicated calculations on a 'computer' that he called the 'difference engine'.

Inspired by these earlier binary technologies, the English mathematician George Boole invented the first complete mathematical formalisation of binary logic in his books *The Mathematical Analysis of Logic* (1847) and *An Investigation of the Laws of Thought* (1854). Boolean algebra demonstrated that one could reduce all of logic and mathematics to a single difference between 1 and 0, plus a few logical operations such as conjunction, disjunction and negation. For Boole, 1 and 0 were not cardinal numbers but merely two different states whose coded combinations represented all possible cardinal numbers. Boole's work was radical but was unapplied until the American mathematician Claude Shannon rediscovered and used it as the basis of a binary code for programming computers with *electrical* circuits.

Electricity is the movement of electrons through the electromagnetic field, which extends in all directions like a vast ocean. When a region of the electromagnetic field interacts with itself, it creates a photon or an electron. Charged electrons absorb and emit photons, causing other electrons to jump between atoms. When a wave of photons is strong enough to move a group of free electrons in the same direction – Einstein's photoelectric effect – this produces an electrical current. It also creates a surplus photon in the form of energy, light and heat. The movement of electrons between the orbits of different atoms creates electricity. Negatively charged electrons flow from negative to positive poles and form an electrical circuit.

Like turning a tap on and off, one can also turn this photoelectric current on or off. When one turns the current on, the technology can complete an electrical circuit that powers various machine parts. One can also use this same current inside a machine as a programming language made up of a series of on or off states. In this way, digital electrical technologies treat the total flow of electromagnetic current as a range of possible electron states that are either flowing, 'on', or not flowing, 'off'.

As I showed in the last chapter, modern scientists understood electricity as an elastic fluid, filled with groups of discrete electrons that expanded and contracted into various forms. Shannon's application of digital logic to electrical fluids fits perfectly into the kinetic paradigm of elasticity in treating electricity as a total state of possible discrete particles. By manipulating the expansion 1 and contraction 0 of this fluid, Shannon was able to create a new ultrafast medium of communication.

This new electrical object was so widespread that it quickly expanded around the world in radio waves and telephone cables, and its language could be contracted and decoded by individual devices. By studying the structure and transmission of binary electrical signals, what Shannon called 'information', scientists and engineers could communicate and store large amounts of information over long distances.

However, as people communicated a lot of information over long distances, there was also a lot of noise. Sometimes a 1 was miscoded as a 0 or vice versa. Shannon called this 'information entropy'. Like all physical systems, electricity is subject to interference and unpredictable movements. Information entropy is the average amount of information produced by a probabilistic data source. Shannon's key idea was to develop several logarithmic equations that defined the range of possible states for a set of binary information. Any set of binary information could be more entropic or less entropic. Shannon's equations treated the electrical field as a *probabilistic* field where specific code patterns were more likely than others. These equations allowed digital communications technologies to effectively determine and filter out what was 'probably' noise and 'probably' signal.

Shannon was the first to figure out a probabilistic coding scheme to overcome the electromagnetic field's fundamentally stochastic nature. As binary digits are produced and transmitted through a material circuit, they necessarily undergo a statistically measurable degree of entropy. This is a beautiful example of how the materiality and movement of matter affects logic. It is also a perfect example of how potential objects respond to the indeterminacy of matter with the logic of probability.

The transistor

The use of digital logic reached new heights when scientists invented transistors in 1947. A transistor is a composite of two semiconductor materials that release a flow of electrons when something electrically stimulates them. A simple transistor is composed of three pieces of silicon layered like a sandwich. The two layers on the ends are silicon doped with phosphorus, and the layer in the middle is silicon doped with boron. The silicon has four electrons in its outer shell, phosphorus has five, and boron has three. Therefore, the electrons from the transistor's phosphorus side move towards the boron side to fill the electron holes in the silicon lattice.

The phosphorus side is the 'source' and the boron side is the 'drain'. In between the two is an electrical contact called the gate, which applies an electrical charge across the two sides. This gate causes the negatively charged electrons of phosphorus (source) to flow into the positively charged holes in the boron, then to phosphorus on the other side (drain). Afterwards, they flow through a connective circuit back around to the source again. With the application of a small voltage, one can create a circulation of electrons between negative and positive charges.

This process is called 'semi-conductivity'. It allows the transistor to act like a binary switch that can be opened or closed by modulating an electrical voltage. The electrical field is not like a series of levers or gears in tension but more like an elastic cloud of possibilities that can be left either in its expanded state of possibility or contracted into a series of actual bits.

However, sometimes the flow of electrons and photons moves outside its most probable quantum state and 'tunnels' through the gate. This might create a 1 instead of a 0 and introduce an error into the code called 'noise'.

The electromagnetic field is an elastic field that continually expands and contracts across many logic gates. The same unstable mobility that lets electrons leap from atomic shell to atomic shell, creating electricity, also allows them to tunnel and deviate from their probabilistic range of possibilities. Like all logics, digital logic relies on movement. Digital logic depends on the elastic expansion and contraction of ranges of possible states of electrons and photons caught in motion, like snapshots.

Transcendental mathematics

The second significant elastic science is what I call 'transcendental mathematics'. Transcendental mathematics is the science of the *possible conditions* for objects. Instead of dealing only with actual objects, this science deals primarily with things such as sets, groups or spaces of possible objects. Modern mathematics took two paths towards creating these new kinds of objects: a geometrical one and an arithmetical one. The first emphasised the *elasticity* of the object, while the second emphasised its *discreteness*.

Non-Euclidean geometry

The modern geometrical object started to emerge around the middle of the eighteenth century with 'elastica theory'. Elastica theory studied bent, curved or warped objects, such as springs, loops, buckles and vibrating strings, and their often non-linear differential equations. The idea at the core of this theory, initially developed by the Swiss mathematician Jacob Bernoulli, was that the coordinates for measuring elastic objects might themselves also be elastic or bendable.

In the nineteenth century scientists used elastica theory to explicitly question the assumption that space itself had to be Euclidean, that is, a flat space where parallel lines never meet. Mathematicians took what they learned from studying elastic objects and applied it to space. The idea was that the Euclidean plane was only one kind of space in a broader range of possibly more elastic or bendable spaces. One of the greatest thinkers in modern mathematics, the German mathematician Carl Friedrich Gauss, worked for years on creating a geometrical refutation of Kant's assertion that Euclidean space was 'an inevitable necessity of thought'.[1] Gauss believed that all previous attempts to prove Euclid's fifth, 'parallel' postulate had failed and that the postulate was fundamentally unprovable. More important, however, without it, Gauss believed that it was still possible to have a perfectly coherent and rigorously defined geometrical space, where there were also some new possibilities. One of these new possibilities was that the sum of a triangle's angles could add up to less than 180 degrees.

Gauss did not, however, arrive at a full-fledged version of non-Euclidean geometry, especially in 3D space.

The Russian mathematician Nikolai Ivanovich Lobachevsky published the first non-Euclidean geometry that treated space as a fundamentally elastic surface where traditional geometrical shapes were deformed and bent. In 1829 Lobachevsky published an article called 'On the Principles of Geometry', that marked the official birth of non-Euclidean geometry and gained him the title of 'the Copernicus of geometry'.[2]

In the 1880s the French polymath Henri Poincaré provided the first simple model of this new kind of space, called the 'hyperbolic disc'. Poincaré replaced flat, infinite, Euclidean space with a finite elastic disc where a straight line was 'any arcs of circles that intersect the boundary of the disc at right angles'.[3] He replaced the idea of an infinite line with the postulate that 'Any line segment can be extended indefinitely in either direction.' As a line approaches the limit of the disc, it becomes infinitesimally contracted and asymptotical.

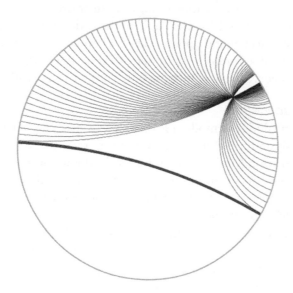

Figure 14.1 Poincaré disc with hyperbolic parallel lines. https://en.wikipedia.org/wiki/Poincaré_disk_model#/media/File:Poincare_disc_hyperbolic_parallel_lines.svg.

Riemann manifolds

The most radical non-Euclidean theory came from Gauss's student, the German mathematician Georg Bernhard Riemann. Riemann proposed that all possible curved metric spaces, not just spheres, were merely exceptional cases of a more general elastic and flexible matrix of space itself, called a 'manifold'. For Riemann, space itself was neither fundamentally Euclidean, disc-shaped, spherical or any other shape. It was a purely elastic topological surface. A manifold was his term for the *conditions* for the possibility of any particular geometric shape *and* any particular background space of any dimensions.

Group theory

Group theory was a mathematical effort to think about spaces and numbers as members of groups that shared structural symmetries. Instead of dealing only with actual numbers and coordinates, group theory, like non-Euclidean geometry, took a more transcendental approach. It studied the structure of possible relationships between elements in a group – their symmetries, operations and invariant features. For example, transformation groups are groups of objects that preserve certain general relationships of an inherent structure even after one rotates, translates or morphs them somehow.

Drawing on the work of the French mathematicians Évariste Galois and Augustin Cauchy on permutation groups and their structural symmetries, the German mathematician Felix Klein unified these ideas into generic structures in which any object might appear, called a 'group'. Groups are potential objects capable of a complete finite set of elastic transformations and relations. In this theory, actual objects are only particular incarnations of groups.

Minkowski spacetime

Another essential idea in modern geometry was that time, like space, was a similarly elastic and non-Euclidean dimension. In 1907 the German mathematician Hermann Minkowski put forward a theory of three-dimensional space that intersected vertically with an expanding and contracting 'time cone' in his paper on 'The Fundamental Equations for Electromagnetic Processes in Moving Bodies'.

Minkowski's idea of spacetime was unique because it was neither Euclidean nor part of Riemann's geometry of manifold spaces. Minkowski did not treat time as a fourth dimension of space but rather as a uniquely elastic dimension of its own, capable of expanding and contracting to different degrees at each point in space. Furthermore, spatial and temporal dimensions did not vary independently from one another but instead changed *relative* to one another.

In this way, Minkowski introduced relativity into transcendental geometry by defining two co-primary elastic fields, space and time. Both expand, contract and deform relative to each other. More than Riemann, Einstein was indebted to his teacher Minkowski for providing a mathematical formalism of a purely elastic and relative transcendental spacetime.[4] In this profound way, Minkowski pushed the mathematical idea of an elastic

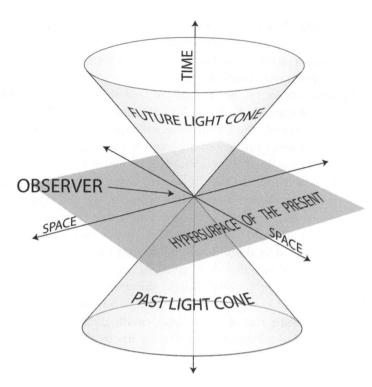

Figure 14.2 Minkowski spacetime. https://en.wikipedia.org/wiki/Minkowski_space#/media/File:World_line.svg.

object to its absolute limits by temporalising space and spatialising time in the same potential object.

Probability and sets

The second mathematical path towards creating new potential objects was more arithmetical, focusing on numbers rather than shapes. Probability theory and set theory developed this path by studying generic sets of possible objects as the conditions for particular ordinal, cardinal or intensive objects. Although probability ranges and sets sound like incredibly abstract objects, I hope to show that they are, like all objects, based on less abstract material and kinetic processes.

Probability

Probability theory had its beginnings in a primarily material and kinetic event: the throw of a die. Modern scientists no longer imagined God as a divine clockmaker but as a dice thrower following a set of hidden laws. The laws of nature, like a die, contained all possible outcomes but were randomly actualised or contracted into a particular series of events.

The Italian polymath Girolamo Cardano wrote the first systematic theory of probability in his book *Liber de ludo aleae* (*Book on Games of Chance*), in 1564. Although the book did not come out until 1663, Cardano's lectures were popular and influenced subsequent generations.[5] In his book, Cardano explicitly postulated the existence of a generic mathematical object which contained a total number of equal alternative possible states; what he called the 'fair die'.[6] Cardano began from the assumption that *all* probability or chance worked like a dice throw. He assumed that his possible object, the 'fair die', had discrete states that remained unaltered before, during and after the throw. Although they sound abstract, there is nothing universal or ideal about potential objects. They have their basis in the particular material fact that dice have discrete sides. Potential objects were formalisations of the act of physically throwing dice without altering their sides and without knowing their outcome in advance. The difference between potential and actual objects is based on the difference between the yet-to-be-thrown and the thrown die. The mathematical model of probability is based on the kinetic act of throwing dice repeatedly without any change in the initial conditions.

Cardano, however, fell short of understanding the potential object. As part of an earlier historical phase associated with tensions and powers, Cardano described dice as having a 'propensity' or power (*possum*) to land in specific ways. Modern probability theorists dismissed such powers in favour of randomness and determinism.

The prevailing modern interpretation of probability was that the world followed fundamentally Newtonian deterministic laws. Given our inability to calculate them for every object, our knowledge only had a certain *probability* of being true. The seventeenth-century French mathematician Blaise Pascal argued that probabilities were to do with how likely our belief was to be true about something objective but unknown, like God's existence. Leibniz similarly imagined probability as degrees of justified belief (between 0 and 1) based on what we know about the world.

This interpretation's key consequence was what the French polymath Pierre-Simon Laplace called 'equipossibility'. In the absence of any specific reason for seeing one outcome as more probable than another, we ought to assume that they are all equally probable, or equipossible.

One regards two events as equally probable when one can see no reason that would make one more probable than the other, because, even though there is an unequal possibility between them, we know not which way, and this uncertainty makes us look on each as if it were as probable as the other.[7]

The theory of equipossibility 'consists in reducing all events of the same kind to a certain number of equally possible cases – that is to say, those such that we are equally undecided about their existence'.[8] Equipossibility, for Leibniz, was not a description of nature but 'an elegant example of *reasoning* about degrees of probability'.[9] Just as it makes sense to treat the outcomes of a die roll as equally probable because we cannot calculate all the physical laws of the throw, so it makes sense to treat nature the same way. But what happens if new events seem to indicate that one event is more probable than another? This was where Jacob Bernoulli's theory of *frequency* came in.

Instead of assuming the equipossibility of events, Bernoulli argued in his posthumous book *Ars Conjectandi* (1713) that we should base our probabilities on how many times particular events occur. As new events occur, we can add those to our ongoing range of possible outcomes and adjust our predictions accordingly. For Bernoulli, the potential object was not a finite totality but a converging set of diverging series. Even if we do not know the complete set of all possibilities, we can still base our predictions on past frequencies as they approach the mathematical limit of an ideal ratio. This is what mathematicians now call the 'weak law of large numbers'.

For example, by flipping a coin repeatedly, we approach the limit ratio of $f(n) \rightarrow 1/2$ for a 'heads' result. What makes this a uniquely modernist theory of the object is that, for Bernoulli, one practically unites the ongoing divergent series into a single convergent ratio (1/2). Instead of continually coordinating divergent series, as the early moderns had, modern mathematicians unified the series into a single *potential* object.

It was not until the English mathematician John Venn wrote *The Logic of Chance* (1866) that someone finally gave Bernoulli's early theory of frequency an 'unambiguous and systematic' treatment.[10] In 1881 Venn drew on George Boole's binary logic to transform Bernoulli's probability theory into a logical system based on whether objects were 'inside' or 'outside' one another. These were his famous Venn diagrams.

By the end of the nineteenth century and into the twentieth, frequentism completely dominated modern probability theory.[11] Although different, all these modern theories shared a common interpretation of the potential object as dice-like. Potential objects were like giant multi-sided dice. One can begin with an assumption about how many sides these objects have or roll them and keep track of their frequency ranges. The same kinetic structure is at play. Both approaches assume that objects are discrete events and that a probability range is a complete expanded range of possible discrete objects.

Sets

Following on the heels of probability theory, set theory also followed the model of the dice throw. Like a die, a set is a discrete, self-identical totality with a subset of discrete sides. Set theory is another abstract formalisation of this concrete kinetic model.[12] However, many mathematicians saw the infinity of natural numbers as a significant barrier to this new model. How could a set ever capture an infinity of numbers?

In 1817 the Bohemian mathematician Bernhard Bolzano proposed a solution. He showed that there were just as many real numbers between 0 and 1 as between 0 and 2 and just as many points in a line segment 1 inch long as in a line segment 2 inches long: namely ∞.[13] If one could make a one-to-one correspondence between every number between 0 and 1 and between 0 and 2, then the two groups of numbers were equally infinite and could be grouped into a single object he called a '*menge*', meaning 'a quantity, multitude, crowd, or aggregate'. Mathematicians later anglicised the term as 'set'. The key idea was that a set was not a number but instead a new kind of object. It was a generic multitude of diverse unspecified elements mixed into a single formal aggregate. Other mathematicians did not accept Bolzano's move to coordinate the 'many' of infinity to the 'one' of a new potential object at the time, and so others had to rediscover many of his results later.

However, if every infinite set contains an infinity of infinite subsets, how do we know that these sets form a continuous number line without leaving any numbers out? How do we know that a set can capture a whole range of numbers without missing any? As early as the 1860s, the German mathematician Richard Dedekind took up this problem and wanted to prove that such an 'arithmetical continuum' existed. His conclusion was as brilliant as it was counter-intuitive. If we select a single point in a line segment, we can divide the line into two parts: everything left and everything right of the point.[14] There is only one point on the line that can bring about this same division. Everything else on the line is either greater than or less than this point. There is nothing left over. 'By this commonplace remark, the secret of continuity is to be revealed', Dedekind wrote.[15]

The secret of the line's geometric continuity, for Dedekind, was that it was not genuinely continuous at all. The only continuum was the 'continuity' of discrete self-identical numbers. If we 'cut' (*schnitt*) our number line at 1, for example, we create two infinite sets. To the left is the set of numbers less than 1, A, and to the right is the set of numbers greater than 1, B. According to Dedekind, between set A and set B there will be no possible number other than 1. We can do this for every point or number on the line, showing the series to be arithmetically continuous without a remainder.

The rational numbers (integers and fractions) are only a small portion of all real numbers. Therefore, Dedekind proposed that the same cut must be done for all irrational numbers, such as $\sqrt{2}$ and π, as well. 'Whenever, then, we have to do with a cut produced by no rational number, we create a new irrational number, which we regard as completely defined by this cut . . . From now on, therefore, to every definite cut there corresponds a definite rational or irrational number.'[16]

For example, if A is the set of all negative and positive rational numbers whose squares are less than 2, and B is all positive rational numbers whose squares are more than 2, then the cut can only be one irrational number: $\sqrt{2}$. For Dedekind, what gives the mathematical continuum its true elasticity of expanding and contracting nested intervals is the absolute discreteness of the number line at every level. This was a defining feature of modern objects. They were elastic only because they were composed of smaller discrete objects.

However, there was a crucial problem in Dedekind's theory of arithmetical continuum. Not all sets are equally infinite, and Dedekind's close friend Georg Cantor was the first to point this out. In Cantor's terminology, we can assign a set a cardinal number of the one-to-one coordinations with its ordinal elements. For example, the cardinal number of a set with five elements is 5. When Cantor applied this 'bijective' logic to Bolzano's and Dedekind's theories of infinite sets, he discovered something he could hardly believe. One can put all integers, rational numbers and prime numbers into one-to-one coordination with one another. For every number in one, there would always be a number in the other to coordinate it with ad infinitum. In this way, Cantor proved that all these numbers were *equally* infinite. However, he also simultaneously discovered real numbers (numbers with infinite decimal expansions), including irrational numbers, that one cannot match up with any other numbers.[17] There are not enough rational numbers to count all of the irrational numbers. Cantor called these infinite sets of real numbers 'uncountable' because one could not count them using his matching method. He also reasoned that these uncountably infinite sets of real numbers must be larger than all 'countably infinite' sets.

Furthermore, a power set (the set of all different possible subsets) is always larger than its original set. In that case, the power set of an uncountable infinite set must be even larger than its original uncountable infinite set. The power set of that power set would be even larger, and so on in ever-larger uncountable infinite sets. In this way, Cantor aimed to develop an even more radical theory of the arithmetical continuum than Dedekind.

However, Cantor quickly realised a problem with his new continuum. If uncountable infinite sets were larger than infinite sets, but they were also *uncountably larger*, how could one know precisely *how much* larger they were? Were they countably larger or uncountably larger? What if there were sets of numbers between the countable infinite sets and uncountable infinite sets? What if there were gaps? How could we ever know if there was an arithmetical continuum or not? Cantor spent the rest of his life trying to solve this problem but never did. [18]

The axiom of choice

Set theory is a highly abstract and formalist theory, but let's look at two of its material and kinetic foundations. The first foundation is the kinetic *act of coordination* that I have tried to describe throughout this book. Objects do not exist on their own but emerge from processes and patterns of coordination. In their own ways, Bolzano, Dedekind and Cantor all coordinated infinite series of numbers into discrete generic objects they called sets. The idea behind this seemingly impossible action was that as long as one could coordinate any series of elements to a set, there was nothing *in principle* that would prohibit one from performing that coordination to infinity. Humans have been coordinating objects since the Palaeolithic.

Still, the new idea behind sets was to act *as if* one had completed counting an infinite series of objects because it was *logically possible* to do so. Sets are, therefore, potential or possible objects. We do not need to flip a coin an infinite number of times to *act as if* the odds of getting heads is 1/2. A set is an object that converges with its total possible states at infinity. Dedekind, though, imagined this as a purely mental and immaterial activity, as if the act of coordination was instantaneous.

> If we trace exactly, what we do in counting a group or collection of things, we are led to the consideration of the power of the mind to relate thing to thing, to let one thing correspond to another thing, one thing copy another, a capacity in general without which thought is impossible. Upon this one, but absolutely inevitable, foundation the whole science of number must be erected.[19]

This was a powerful new kind of object because it allowed mathematicians to treat even infinitely non-repeating irrational numbers as if they were a single generic discrete object. As Dedekind wrote, 'To every definite cut there corresponds a definite rational or irrational number.'[20]

However, there is no necessary reason why we must accept the legitimacy of this infinite coordination. The German logician Ernst Zermelo argued in 1904 that it was a presupposition or 'axiom' of set theory itself. We can decide to act *as if* the infinite coordination of sets and subsets were possible *without actually doing the coordination*. In 1940 Kurt Gödel proved that this axiom was consistent with set theory's other axioms. However, in 1963 the American mathematician Paul Cohen showed that it was also independent of the other axioms and one could not prove it within the system.[21]

Therefore, the theory of infinite sets relied on a non-mathematical presupposition of an infinite and instantaneous act of coordination. We can also decide to act *as if* there is an arithmetical continuum between countable and uncountable infinities. Cantor called this the 'continuum hypothesis', and it is another unprovable axiom of set theory.[22] In this way, one can add and subtract axioms to set theory to expand or contract its range of possible objects.

Conclusion

This chapter concludes Part II of this book on the history of objects. In it I have tried to trace the historical emergence of four kinds of objects – ordinal, cardinal, intensive and potential – and describe the unique kinetic patterns of each. Although my history was broadly chronological, it was not developmental. Potential objects are not superior to ordinal objects. Furthermore, once scientists invented each type of object, it did not disappear when a new one emerged but persisted alongside it. A theory of objects should seriously consider the deeply historical conditions and the patterns of motion that continue to sustain these objects into the present.

Part III of this book is about the present. I think we can see the emergence of a new theory of the object at the turn of the twenty-first century in recent interpretations of quantum physics, category theory and chaos theory. We can also find the hybrid coexistence of the previous four kinds of objects alongside this new object. Let us see what this new object is and how it can shed new light on the others.

Notes

1. Cited in Leonard Mlodinow, *Euclid's Window: The Story of Geometry from Parallel Lines to Hyperspace* (New York: The Free Press, 2001), 117.
2. Uta C. Merzbach and Carl Boyer, *A History of Mathematics* (Hoboken, NJ: Wiley, 2013), 494. See also Jeremy Gray, *Worlds Out of Nothing: A Course in the History of Geometry in the 19th Century* (London: Springer, 2007).
3. Mlodinow, *Euclid's Window*, 122.
4. Mlodinow, *Euclid's Window*, 128.
5. Cited in Ian Hacking, *The Emergence of Probability* (Cambridge: Cambridge University Press, 2006), 54.
6. Girolamo Cardano, *The Book on Games of Chance: The 16th-Century Treatise on Probability*, trans. Sydney H. Gould (New York: Dover, 2016), ch. 9.
7. Cited in Hacking, *The Emergence of Probability*, 132.

8. Cited in Hacking, *The Emergence of Probability*, 132.
9. Gottfried Leibniz, *Opera Omnia*, ed. Louis Dutens (Geneva, 1768), vol. VI, 318, my italics.
10. Cited in Hacking, *The Emergence of Probability*, 53: 'These evident facts need emphasizing because Carnap said that in 1866 John Venn "was the first to advocate the frequency concept [. . .] unambiguously and systematically" [. . .] Venn the alleged first frequentist, wrote that "the fundamental conception which the reader has to x in his mind as clearly as possible, is, I take it, that of a series. But it is a series of a peculiar kind, one of which no better compendious description can be given than that which is given by the statement that it combines individual irregularity with aggregate regularity." We have no doubt that Cardano had fixed this idea in his mind.'
11. For a definition and interpretation of frequentism, see David Howie, *Interpreting Probability: Controversies and Developments in the Early Twentieth Century* (Cambridge: Cambridge University Press, 2007).
12. For a more complete history, see Jose Ferreiros and Domâinguez J. Ferreirâos, *Labyrinth of Thought: A History of Set Theory and Its Role in Modern Mathematics* (Basel: Birkhäuser Basel, 2007); and Merzbach and Boyer, *A History of Mathematics*.
13. Merzbach and Boyer, *A History of Mathematics*, 457.
14. Merzbach and Boyer, *A History of Mathematics*, 536.
15. Cited in Merzbach and Boyer, *A History of Mathematics*, 537.
16. Richard Dedekind, *Essays on the Theory of Numbers: I. Continuity and Irrational Numbers, II. The Nature and Meaning of Numbers*, trans. Wooster Woodruff Beman (Chicago: Open Court, 1901), 15.
17. See Ferreiros and Ferreirâos, *Labyrinth of Thought*; and Merzbach and Boyer, *A History of Mathematics*, 541.
18. Charles Sanders Peirce also adopted a Cantorian theory of the continuum. In his later work, however, he started to develop a much more 'vague' and purely 'conceptual idea' of the continuum as a way of avoiding the analytic problem posed by Cantor's hypothesis. However, in this version of the continuum we can see that Peirce has merely hypostatised the ahistorical idea of an immaterial realm of 'pure generic possibility' beyond all 'actual multitudes' or existing 'individuals'. Peirce's continuum is thus a 'homogeneous' and undifferentiated wholeness. For an excellent summary of this more conceptual-idealist version of the continuum, see Fernando Zalamea, *Peirce's Continuum: A Methodological and Mathematical Approach*, available at <http://uberty.org/wp-content/uploads/2015/07/Zalamea-Peirces-Continuum.pdf> (last accessed 4 March 2021). 'The idea of a general involves the idea of possible variations which no multitude of existent things could exhaust but would leave between any two not merely many possibilities, but possibilities absolutely beyond all multitude.' Charles S. Peirce, 'Lectures on Pragmatism', in *Collected Papers of Charles Sanders Peirce, Volume V: Pragmatism and Pragmaticism*, ed. Charles Hartshorne and Paul Weiss (Cambridge, MA: Belknap Press of Harvard University Press, 1931), 103. 'Generality is, indeed, an indispensable ingredient of reality; for

mere individual existence or actuality without any regularity whatever is a nullity. Chaos is pure nothing.' Peirce, 'What Pragmatism Is', in *Collected Papers*, vol. 5, 431.

19. Richard Dedekind, *What Are Numbers and What Should They Be?*, ed. H. A. Pogorzelski, W. J. Ryan and W. Snyder (Orono, ME: Research Institute for Mathematics, 1995), viii. Cited in Ernst Cassirer, *Substance and Function and Einstein's Theory of Relativity* (Mineola, NY: Dover, 2003), 36.

20. The modern German philosopher Ernst Cassirer argued that mathematics was indebted to Immanuel Kant's theory of transcendental temporality. Because Kant said that the lived time of the subject was structured as a progressive series of distinct and discrete moments (before, now, after), the human mind, according to Cassirer, was capable of using this structure for the ordering of generic arithmetic units. The mind performs a many-to-one coordination of the manifold into a unitary 'I' 'now'. On William Hamilton's definition of algebra as 'science of pure time or order in progression' and its relation to the Kantian concept of time, see Ernst Cassirer, 'Kant und die moderne Mathematik. (Mit Bezug auf Bertrand Russells und Louis Couturats Werke über die Prinzipien der Mathematik.)', *Kant-studien*, 12 (1907), 1–49 (34). 'The intuition of pure time, upon which Kant based the concept of number . . . Arithmetic can be defined as the science of pure time only when we remove from the concept of time (as Hamilton does, for instance), all special determination of character, and merely retain the moment of "order in progression".' Cassirer, *Substance and Function*, 40.

21. Merzbach and Boyer, *A History of Mathematics*, 560.

22. Dedekind's proof rests implicitly upon the axiom of choice. See Ferreiros and Ferreirâos, *Labyrinth of Thought*, 237.

PART III

THE CONTEMPORARY OBJECT

V THE LOOP OBJECT

15

The Pedetic Object

What is the prevailing object of our time? Unlike previous kinds of objects, contemporary objects do not have a single pattern of motion. In Part III of this book, I argue that they are *hybrid* objects that mix the four primary kinds of objects and kinetic patterns I described in Part II. They can do this because they are also *indeterminate* objects.

To understand the contemporary objects of our time, we need to see how they mobilise and mix the previous objects. This book has tried to show as concretely and historically as possible that there is not just one kind of object. Instead, there are several prevailing *processes or patterns* that converge and diverge through history. Science does not discover pre-existing objects but co-creates them and then uses them to reorganise the world of things. This can help us survive and make life easier but it can also be dangerous and self-destructive.

There is something unique about the present historical moment and the nature of contemporary objects that lets us see the indeterminate and hybrid nature of all prior objects. In Part III of this book, I want to trace the rise of a new kind of object that, in my interpretation, brings to the foreground of our attention three hidden dimensions of all previous objects: hybridity, pedesis and feedback. These dimensions have always been part of objects but were either ignored or seen as problems to be solved.

What is interesting to me about quantum theory, category theory and chaos theory is how they confront these three aspects more directly than previous sciences. In the chapters that follow, I want to offer a movement-oriented interpretation of these contemporary sciences and see what this means for the future of objects more generally. I chose to look at these three sciences because each claims to be a foundational science in its own way. Indeed, these sciences have had significant consequences for almost every field of knowledge in the early twenty-first century. What distinguishes these sciences from others, such as biology, neuroscience, geology, chemistry and economics, is their aim to describe *all objects*. It is, therefore, no surprise that philosophers have taken such interest in them.[1]

Quantum theory, category theory and chaos theory were all born in the twentieth century, but they did not attain their most robust formulations, influence and application until the twenty-first century. As a philosopher, I aim to offer synthetic and interpretive concepts that help make sense of a wide range of diverse practices and knowledge. In the following chapters, I propose a synthetic and interpretive idea for a new kind of object that I call the 'loop object'. This concept aims to identify three common methods and a new pattern of motion shared by these three sciences.

Specifically, I argue that the objects studied by these three sciences all directly respond to the phenomenon of indeterminate motion and its resistance to discrete or static objectification. Yet, simultaneously, indeterminacy plays a generative role as well. Or at least this is what I would like to argue in my interpretation. The name I give to this concept of generative resistance at the heart of contemporary science is the 'kinetic operator'. In these chapters, I want to show that there is another way to interpret what these sciences are doing than the interpretations currently offered.

Quantum theory, category theory and chaos theory highlight the phenomena of hybridity, pedesis and feedback to some degree and, in my reading, reveal some of the non-objective conditions for objects. In my view, these features distinguish them from ordinal, cardinal, intensive and potential objects.

Therefore, there are two goals for Part III of this book. First, I want to point to all the places where these three sciences have made historically significant inroads towards developing a new theory of objects. Second, I want to show how each science directly grapples with the phenomenon of indeterminate movement, which supports and resists objectification. My aim here is not to make any new predictive claims about what these sciences will discover in the future. Instead, I want to offer a kinetic interpretation of what they are doing, what they share in common, and what this tells us about our historical moment.

I have divided the following chapters into four sections each. The first three describe the hybridity, pedesis and feedback of the science, and the fourth section identifies the role of the kinetic operator. In this introductory chapter, I want to introduce and define the loop object and its four features.

The loop object

The loop object is the name I give to the pattern of motion created by the three contemporary sciences discussed in these chapters. I offer here a synthetic interpretation of these sciences that is consistent with the kinetic theory of the object that I have developed in this book. There is no theoretical consensus on interpreting these sciences, but I would like to propose my own in these chapters. Typically, each of these sciences interprets only its own practice, but here I would like to offer a broader interpretation that makes sense of what is going on in more than one scientific field. This has been my method throughout this book. Here I apply it to the present. I hope to show that my kinetic theory of the object is not only consistent with recent developments in science but that it may help us see the emergence of a new kind of object.

I call this new kind of object a 'loop' object because of the way these three sciences draw attention to the *folded*, *iterative* and *metastable* nature of objects. Loop objects do not come pre-made, but someone or something has to sustain them through an ongoing and indeterminate process. In my interpretation, loop objects show us that the conditions of objects are not objects, but *processes*. This novel thesis is at the core of this book more broadly.

If this thesis is correct, it has significant consequences for our understanding of all objects. In the interpretation I present here, the contemporary loop object brings to the historical foreground a spectre that haunts all previous objects: the non–objective material process of folding, looping and kinetic coordination at the heart of objects.

I define the loop object by four features. Let's briefly look at each of these before moving on to a much closer study of them in quantum, category and chaos theories.

Hybridity

The loop object is a hybrid mixture of previous objects because it provides a theoretical framework that accounts for all earlier kinds of objects. For example, older objects found in classical mechanics, algebra or dynamics are not excluded or negated by the loop object but reinterpreted within a larger framework.

Pedesis

The loop object is the metastable result of an indeterminate process. In contrast to previous kinds of objects, the loop object does not treat objects as discrete and static. Instead, I understand them as emergent properties of processes. In my interpretation of quantum, category and chaos theories, objects are not things that occur as actualisations of fixed ranges of possible objects. Loop objects are neither deterministic, random nor probabilistic. They are metastable states of fundamentally indeterminate processes.

Feedback

Loop objects are also relational and formed by feedback. When material processes react to or affect themselves iteratively, they form 'loops'. The object's metastable self-identity or unity results from a continual self-interaction or feedback with itself and others. Objects and their relations are co-emergent phenomena. One does not come before the other, but rather both emerge at the same time. Feedback does not occur between two discrete individuals – individual objects co-emerge from the process of feedback.

These are the three features I want to show at work in contemporary science. In the following chapters, my aim is not to provide an exhaustive history of quantum, category and chaos theories, but to provide just enough context and exposition to clarify my interpretation. What follows is not a literature review but an attempt at a unique kinetic understanding of these sciences. I want to highlight what are, in my view, the most significant and novel aspects of these sciences compared to other historical kinds of objects. Along the way, I will try to point out precisely where my interpretation diverges from others.

The kinetic operator

Each of the following chapters concludes with a description of how each science responds to and works with processes of indeterminate movement. The loop object is such a difficult kind of object to study because it lacks a static foundation and is not entirely discrete. These two features were critical aspects of previous historical theories. Nonetheless, I want to argue that these three sciences, in my interpretation, yield critical new insights about the fundamentally unstable and indeterminate nature of motion at the heart of all objects.

Twenty-first-century science is at an incredible historical conjuncture where a new theory of objects seems possible. If they make any assertions at all about the nature of things, most working scientists are still committed to a modernist interpretation of objects as potential or probable ranges of discrete states. Others treat the world as potential objects but remain agnostic about whether this says anything about the nature of things.

My argument is that there is a third way. What if nature was neither deterministic nor random but *relationally indeterminate*? What if the conditions of objects were not smaller discrete objects but *processes*? I do not mean to suggest that I know for sure that nature is a relationally indeterminate process forever and all time, but only that such an interpretation is at least consistent with what we know at this point. Just because scientists use statistical or deterministic equations to anticipate how matter might move does not mean that matter *is* statistical or deterministic. It only means that certain equations give us a good guess about what might happen. However, since no current equation is 100 per cent accurate or complete, this leaves the door open for alternative interpretations.

This is what I want to show in the following chapters. Perhaps some scientific novelty will prove my interpretation inconsistent, but until then, I think it's fair to propose an alternative understanding that fits what we currently know.

Let me clarify what this third interpretation is and how it differs from three aspects of the modern one. The first aspect of the modern theory of objects is that nature is entirely deterministic. The world only appears to be random and unpredictable because we don't have the right equations yet or cannot practically apply them to all the variables. Perhaps there are many deterministic universes, and we don't yet know which one we are in. God, as Einstein said, plays dice all the time under deterministic rules he created. Maybe one day determinism will be proven correct, but since we do not have any equations that are 100 per cent predictively accurate or observer-independent, determinism remains inconsistent with present experience. The theory that there are many deterministic worlds, though, is neither provable nor disprovable.

Another aspect of modern objects is randomness. If any movement of matter in the universe were genuinely random, it would have to be unaffected by anything else. Nature would have to be globally non-relational. However, this idea is inconsistent with recent experimental evidence of non-local quantum entanglement, which we will discuss in the next chapter. God does not throw random dice.

A third aspect of the modern object is its probabilistic interpretation. This comes in two flavours. One can use probability theory because it offers a good guess about what things may do, and one can remain agnostic about the nature of things. Alternatively, one may genuinely believe that nature is deterministic or random, but since we have no absolute proof of that yet, one may use probability models as ways of approximating this. My worry in the first case is that science ultimately gives up on reality and assumes that probability distributions and human observation can have nothing to with how nature really moves. This position may also violate the phenomenon of observer-dependence and entanglement. It is an anti-realist position. My worry about the second case is the same as those about determinism and randomness described above.

The alternative interpretation I propose in the following chapters is a *historically realist* one. Given what we know, nature seems to be a relationally indeterminate process. Or at least this interpretation is consistent with what we currently know and does not posit any higher metaphysical explanation. Science has always sought explanations for the movements of matter, but what if there are none? What if matter's movements are neither deterministic nor random, not fully compatible with any range of possible states we may imagine? If matter's movement were a relationally indeterministic process, then the world would look a lot like it does to us today. It would be in constant motion, non-locally entangled, but would not be fully predictable by our best equations.

I am not saying we should altogether jettison the modern theory of the object. In the history of objects, all the old fields persist, but in a transformed way. We can keep certain aspects of probability theory but interpret them in a new way that does not commit us to any metaphysical belief in the determinism, discreteness or randomness of the universe. One cannot deny how effective statistical methods have been in the sciences. The question is why. I do not think we need to assume that this success is due to ontological determinism or randomness.

In my kinetic interpretation, objects and their fields are metastable states of indeterminate relational processes. Probability fields are no different. We can study events without assuming they are discrete or independent of our observations. We can study frequencies of events without believing there is a prior totality of them. We can study frequencies without assuming there is a fixed rate or a random distribution.

This would change the meaning of what many scientists call probability, but I think there are several benefits to changing our interpretation. First of all, it would actively raise awareness that when we make objects, those objects are not entirely discrete. They are related to our process of making them in some way, no matter how seemingly trivial. What often seems trivial is what is most important in the creation of new knowledge. If we treated objects as relational processes, we might find new connections to the natural world that we might have bracketed out before. If we treat objects as fundamentally unstable and indeterminate, it might cause us to pay more attention to the singularity of the event of observation. If we treat objects as indeterminate processes lacking a totality of possible states, it might encourage us to be more sensitive to ongoing singular changes. This kinetic interpretation might also more accurately reflect the genuine novelty and creativity of matter in motion and scientists' role as co-producers of that novelty. They are not neutral observers but *kinetic operators*.

This philosophical interpretation does not aim to change any particular thing about what scientists are doing, but how they understand what they are doing and give a name to the non-objective conditions of the objects they study. There are no objects without a constitutive kinetic process of entangled measurement *immanent to them*.

In the next three chapters, I want to look at quantum, category and chaos theories from this perspective and find the kinetic operator in each.

Note

1. For example, Alain Badiou, *Mathematics of the Transcendental*, trans. A. J. Bartlett and Alex Ling (London: Bloomsbury, 2017); Manuel DeLanda, *Intensive Science and Virtual Philosophy* (New York: Bloomsbury, 2013); Karen Barad, *Meeting the Universe Halfway: Quantum Physics and the Entanglement of Matter and Meaning* (Durham, NC: Duke University Press, 2007); Rocco Gangle, *Diagrammatic Immanence: Category Theory and Philosophy* (Edinburgh: Edinburgh University Press, 2016); Vicki Kirby, *Quantum Anthropologies: Life at Large* (Durham, NC: Duke University Press, 2011); Ilya Prigogine and Isabelle Stengers, *Order Out of Chaos: Man's New Dialogue with Nature* (London: Verso, 2017); Michel Serres, *The Birth of Physics* (New York: Rowman and Littlefield International, 2018).

16

The Contemporary Object I: Quantum Theory

The first loop theory of the object was born in the 1920s in quantum field theory. Quantum field theorists wanted to know where quanta come from and how they move. However, modern quantum theories in the early twentieth century were still non-relativistic. They could not describe how electrons emitted and absorbed photons at the speed of light as they moved between energy levels.

In 1928 the English theoretical physicist Paul Dirac, whom many consider the greatest physicist of the twentieth century after Einstein, struck upon an ingenious idea. Instead of treating objects as discrete and static quanta with fixed properties, what if we treated them as *relationally constituted*? What if position, velocity, angular momentum and electrical potential only emerged through *interaction with other objects*. Then we would have a fully relativistic quantum theory to account for quanta-in-transition.

Shortly after Dirac formalised this idea mathematically, he realised that it unified two previously distinct kinds of objects: fields and particles. Classical particles were supposed to be inelastic, and fields were supposed to be elastic. Physicists still thought of particle-wave duality in terms of classical objects. They thought of particles and particle waves moving through homogeneous space and time.

However, when Dirac mathematically unified fields and particles, he followed the model proposed by special relativity where space and time were not homogeneous. For Dirac, a quantum field is different from a classical one because it was not a set of fixed topological points. For Dirac, a quantum field acts instead like a vibrating guitar string whose frequencies or excitations give rise to the appearance of particles at discrete levels or quanta of energy. Dirac treated particles and fields as vibrations in the same moving matter. A photon, he argued, was an excitation or vibration in a continuous and indeterminate electromagnetic field. He called this 'quantum electrodynamics'.

Dirac's equations did not solve everything, but they were successful enough approximations to help the Japanese physicist Sin-Itiro Tomonaga and others in the 1950s[1] explain how different charged particles might also emerge and interact with one another as relativistic fields.

In this chapter, I want to show how the study of quantum fields changed our theory of the object and opened the door to a more hybrid, pedetic and relational loop object.

Hybridity

Quantum field theory is a hybrid scientific framework unifying most areas of physics. After completing quantum electrodynamics in the 1950s, quantum field theory was extended over the next twenty years to explain strong and weak nuclear forces. This resulted in the unification of all three fields in a single new framework called 'spontaneous symmetry breaking' in 1962. The idea was that all three fields were part of an original field that split into others at some point early in the universe. Theorists speculated that there must have been a more fundamental symmetrical field that no one had yet observed. They called it the Higgs field.

By the 1970s physicists had unified all the observed fields, particles and forces, except gravity and the strong force, in the single most successful model of fundamental physics ever devised, and dubbed it 'the standard model'. Scientists completed this model in 1973, and it has held up ever since in a wide array of experiments.[2] Today we know of fifteen quantum fields, whose quanta are the elementary particles, including electrons, quarks, muons, neutrinos and Higgs bosons.[3] 'Today, all theories of elementary particles (such as the quark theory of matter) are quantum field theories. Particles are thought of as energetic excitations of the underlying field.'[4]

From the 1970s to the present most discoveries have merely confirmed and elaborated on the standard model. This includes the discovery of the Higgs field in 2013 and of gravitational waves in 2017. Today, quantum field theory remains the dominant and prevailing physical theory of reality, and its equations have significant applications in computer engineering, chemistry, molecular biology and other areas.

Physicists are not shy about this accomplishment. As the American theoretical physicist Lee Smolin says, 'It describes almost all we see, with the exception of gravity.'[5] The English particle physicist Frank Close writes that '[Dirac's equation] is today recognised as the seed of everything that underpins chemistry, biology, and in principle life itself.'[6] And the American physicist Sean Carroll describes Dirac's equations as the 'equations underlying you and me and the world of our everyday experience'.[7] The standard model equations fit on a T-shirt and describe all of physical reality – with the single exception of gravity.

This is where the most recent addition to quantum field theory comes in. Quantum gravity theory uses field theory equations to describe the emergence of space and gravity from quantum fluctuations of energy. The standard model does not yet unify quantum field theory and general relativity. However, an increasing number of theoretical physicists are currently working on this cutting-edge field and have developed several internally coherent mathematical formalisations assembled from existing equations.

Although some quantum theories of gravity are, in principle, experimentally testable, the conditions for such tests are challenging to obtain. Measurements from tiny fluctuations in the cosmic background radiation or the explosions of black holes might test these theories.[8] Quantum gravity still awaits confirmation, but for many physicists it is the most likely candidate for a future unified 'theory of everything'.[9]

The standard model is not just a new theory that leaves others behind; it is a *hybrid* or meta-theory that compiles all the pieces of the most experimentally successful theories in physics into a single hybrid equation that results in finite answers. Even though it has none of the elegance and simplicity of Einstein's relativity theories, the standard model still gets the job done. It has astounding experimental accuracy across an impressive number of scientific fields as long as we deal with finite constants and discrete results. This is what allows for the universal *hybridity* of the standard model of quantum field theory.

Pedesis

Quantum field theory is also indeterministic. What is the energy of a field that has not produced a discretely measurable particle? When quantum field theorists tried to answer this question, they found that the lowest energy of a field was not zero or any *determinate* amount at all. Quantum fields have tiny *indeterminate* vibrations called 'vacuum fluctuations'. These fluctuating states are neither in one state nor another and so are not technically objects.

Quantum fields never stop indeterminately moving even when in stable observable particles. Although we associate the term 'vacuum' with void or emptiness, this is not the case. 'A quantum field implies that its vacuum is a humming hive of activity. Fluctuations continually take place, in the course of which transient "particles" appear and disappear. A quantum vacuum is more like a plenum than like empty space.'[10]

Particles are not floating in a void. As the physicist Frank Close writes,

A vacuum is not empty but seethes with transient particles of matter and antimatter, which bubble in and out of existence. Although
these will-o'-the-wisps are invisible to our normal senses, they dis-
turb the photon and electron in the moment of their union and
contribute to the number that the experiment measures.[11]

Physicists call these transient particles 'virtual particles' even though they
are neither virtual nor particles but are *real indeterminate kinetic vibrations* in
the field itself.

Physicists describe the motion of these fluctuations as 'turbulent whirlpools' and their effect on the equations of quantum field theory as 'perturbation theories'. But vacuum fluctuations do not merely 'perturb' particles.
Particles are vibrations in the fields. In quantum field theory, all matter is
fluctuations that originate from the vacuum at some point. Indeed, most of
the mass of protons and neutrons is not due to their quarks, which make
up only 1 per cent of their mass, but the result of the *movements* of their
indeterminate vacuum fluctuations or 'virtual particles'.[12] Nature does not
abhor a vacuum; it *adores a vacuum.*

However, the existence of indeterminate vacuum fluctuations poses an
interesting problem. If science's role is to create and order objects, what
is its relationship to non-objective fluctuations of energy? If the equations
of quantum field theory treat energy as a relationally relativistic[13] process,[14]
without any locally hidden deterministic variables,[15] where do the particles
come from? If one dealt with discrete particles or even particle-waves,
then there would be only a finite number of possible outcomes in a closed
system. However, if one is dealing with *indeterminate fields*, there is a much
broader range of 'degrees of freedom'.[16]

Once you try to study a process with indeterminate degrees of freedom,
the whole probabilistic model starts to breaks down. Without finite discrete
states to choose from, there is no probabilistic prediction. If one wants to
predict a finite answer, one must treat the field as a set of possible discrete
objects. Even then, the chance of one of these objects happening is still $1/\infty$
or 'infinitely approaching, but not equal to, zero'. Ultimately, if one wants a
prediction of probability, one also has to select a finite range of discrete determinate states. Therefore, to generate finite practical predictions about how an
indeterminate field might move, quantum field theorists have to treat fields *as
if* they were finite probability ranges of discrete particles moving randomly.[17]

In my view, this does not necessarily mean that indeterminate field fluctuations are *genuinely* random, discontinuous or continuous. If they *are* indeterminate, they cannot *be* anything determinate at all. If they are neither

substances nor particles, they cannot be continuous or discrete. The standard model equations merely mean that if physicists want probabilistic predictions, they will have to act on an indeterminate field by measuring only a finite range of finite possible states. Since the action of measuring an indeterminate field introduces energy, scientists change the field by measuring it. Scientists cannot make objective measurements of quantum fields. They are co-creating and stabilising an indeterminate process into an object. The indeterminate movements of the field never go away completely. They just become meta-stabilised into objects.[18]

Feedback

The third feature of quantum field theory is that its objects form through intra-active and relational mechanisms, or what I call 'feedback'. This process is different than in the modernist theory, where pre-established discrete particles interact with one another following deterministic or statistical laws. Discrete particles *interact*, but fluctuating fields *intra-act*.[19] If fields are not discrete objects, they do not relate as absolutely different things but rather as dimensions of the same process.

In quantum field theory this is called a 'backreaction'. A backreaction is when a field keeps responding to itself indefinitely and makes it difficult for scientists to measure the discrete mass of a particle. 'One important way in which this happened was through interaction with the restlessly fluctuating vacuum', as the physicist John Polkinghorne notes.[20] As the ambient vacuum fluctuates and responds to itself like waves of water in a moving tub, it perturbs the metastable state of its particles.

These backreactions were one of the reasons Dirac was never fully satisfied with his equations. If fields are relational *and* continually changing, then each change keeps changing all the field relations repeatedly.[21] This is also why Dirac rejected Schrödinger's wave equation as a superposition of possible deterministic paths a particle might take.[22] 'The reality of the electron is not a wave: it is how it manifests itself in interactions', Dirac said.[23]

Loops

One can visualise the immanent feedback relations in quantum fields as loops, folds or bubbles. Excitations in these field lines produce a 'vortical whirling'[24] or 'bubbling'[25] effect on the surface of the field, causing it to interact with itself and other fields. The field lines 'loop' back over themselves in cycles of continual feedback and transformation.

This is the basis of what is called loop quantum gravity theory. Loop quantum gravity treats these loops as composing the primarily 'granular'

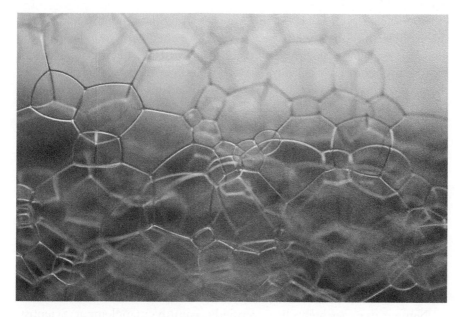

Figure 16.1 The foam of bubbles. https://commons.wikimedia.org/wiki/File:
Foam_Bubbles.jpg.

texture of spacetime itself.[26] The assemblage of all these active quantum
field loops produces what physicists call a 'spin foam network'. They call it
that because the loops weave together to create the foam-like dimensional-
ity of space. The Italian physicist Carlo Rovelli calls it '"Foam" because
it is made of surfaces that meet on lines, which in turn meet on vertices,
resembling a foam of soap bubbles.'[27]

In this quantum gravity theory, objects are products of processes of
folding. As Rovelli explains,

> In the world described by quantum mechanics, there is no reality except
> in the relations between physical systems. It isn't things that enter into
> relations, but rather relations that ground the notion of things. The
> world of quantum mechanics is not a world of objects: it is a world
> of events. Things are built by the happening of elementary events. As
> the philosopher Nelson Goodman wrote in the 1950s, with a beauti-
> ful phrase: 'An object is a monotonous process.' A stone is a vibration
> of quanta that maintains its structure for a while, just as a marine wave
> maintains its identity for a while, before melting again into the sea.[28]

In other words, according to Rovelli, the conditions of objects, including spacetime, are not other objects but *relational processes*. Objects are

emergent features of fields. As the American physicist Karen Barad writes, the particle 'intra-acts with itself (and with other particles) through the mediated exchange of virtual particles . . . [thus] the energy-mass of this infinite number of virtual intra-actions makes an infinite contribution to the mass of the electron'.[29] In this way, the quantum theory of the object is made from an indeterminate feedback effect.[30]

Entanglement

Another crucial instance of quantum feedback is entanglement.[31] In Schrödinger's words, entanglement is '*not* . . . *one* but rather *the* characteristic trait of quantum mechanics, the one that enforces its entire departure from classical lines of thought'.[32] Quantum entanglement occurs when we cannot describe the state of a pair or group of particles independently of the state of the others it is entangled with, even when a considerable distance separates them. When we measure the position, momentum, spin and polarisation of one of the entangled particles, it correlates with another.

Numerous experiments have verified quantum entanglement. Scientists now accept that it is not possible to specify a system's state by listing the state of all its subsystems individually. 'We have to look at the system as a whole, because different parts of it can be entangled with one another.'[33] But how big is the whole system?

In 1935 Albert Einstein, Boris Podolsky and Nathan Rosen (EPR) argued that if entanglement were real, then either of two things followed. Either quantum physics was incomplete because it could not explain entanglement, or it implied a non-local 'spooky action at a distance' (a phrase Einstein later came up with). Neither option was acceptable for EPR. Einstein believed that objects should have fixed properties whether or not one could measure them. He wrote,

> That which really exists in B should not depend on what kind of measurement is carried out in part of space A; it should also be independent of whether or not any measurement at all is carried out in space A. If one adheres to this program, one can hardly consider the quantum-theoretical description as a complete representation of the physically real. If one tries to do so in spite of this, one has to assume that the physically real in B suffers a sudden change as a result of a measurement in A. My instinct for physics bristles at this.[34]

The so-called 'EPR paradox' remained a theoretical controversy until the Irish physicist John Bell, inspired by the American scientist David Bohm's earlier non-local hidden-variable theory (1952),[35] put forward a theorem and thought experiment that proved Einstein wrong (1964).[36] However, the French physicist Alan Aspect's 1982 EPR-based experiments in Paris proved entanglement undeniably real. Over the past forty years, scientific experiments have repeatedly confirmed statistical correlations between distant quantum events. All theories that have tried to explain entanglement by local causal factors have been proven false.[37]

The consequences of quantum entanglement are radical, but there is no universally accepted interpretation for their cause. We know that local motions are entangled or statistically correlated with global ones, but this does not necessarily entail any action at a distance or retroaction in time. Nor does it necessarily follow a deterministic wave function for the entire universe, as David Bohm had argued.[38]

In my kinetic interpretation, we do not necessarily need to explain entanglement by any broader causal framework. Indeed, it may even be impossible to do so because of the indeterministic nature of quantum fields. Perhaps indeterminate entangled correlations are the givens of a new theory of the object without a higher causal explanation.[39] Each local change occurs in feedback with the global indeterministic universe. In this way, there is a 'qualitative transformation of the whole', as Bergson was fond of saying.[40]

Hybridity, pedesis and feedback, or as Carlo Rovelli calls them, 'granularity, indeterminacy, and relationally', are the three aspects of quantum field theory that, in my view, contribute to a new loop theory of the object.[41]

The kinetic operator

In this next section, I want to show the role that indeterminate movement, or what I call the 'kinetic operator', plays in creating quantum kinetic objects. The kinetic operator plays a crucial role in renormalisation, the practice of bracketing out the indeterminacy and feedback in quantum fields to get finite predictions for mathematical equations. As one of Dirac's students, Frank Close, writes, 'The philosophy of renormalisation is one of the most difficult, and controversial, in all of particle physics. Yet it underpins modern theory.'[42]

Since the mid-1930s quantum field theorists have been highly aware of, and often frustrated by, this limitation. Even Dirac was aware that his equations treated electrons as if they did not act on themselves in 'backreactions' or 'self-energies' and as if they moved through empty space. When one applies Dirac's equations rigorously to an electron's energy, it results in an infinite mass, which is experimentally impossible. The same thing happens if one uses his equations on the empty space around the electron, the so-called 'quantum vacuum'.

This 'infinity problem' was solved in the 1940s when, among other things, physicists started taking initial measurements of the charge and mass of an electron.[43] One cannot use quantum field theory to calculate the electron's charge or mass from theory alone without getting infinite answers. Therefore, scientists used experimental measurements as 'benchmarks for anything else that we may wish to compute . . . relative to these experimentally determined quantities'. 'The marvel', Close writes, 'is that instead of infinity, all the answers now turn out to be finite, and, even better, the values are correct.'[44] This was an immensely successful innovation, making quantum electrodynamics the best experimentally confirmed theory ever.

However, making an initial experimental measurement does not eliminate indeterminacy, vacuum fluctuations or feedback reactions. Instead, it acts as a *kinetic operator*. An act of measurement requires that scientists shoot photons at something using a measuring apparatus. However, these photons and their experimental set-up have their own indeterminate quantum fields. Therefore, measurement is more like an intra-action of fields that actively shapes and co-determines the outcome while simultaneously leaving something out.

The American physicist Richard Feynman realised that measurement always left something out. He developed a hierarchical method for diagramming Dirac's equations into distinct possible path equations at simpler and more complex levels of how particles might interact between the first and final measurement points. His practical strategy was to calculate the simplest possible interactions first and temporarily ignore the others.[45] To get a closer approximation, one could always do more and more calculations to refine the approximation.[46] Ultimately, though, one can never get to the bottom.

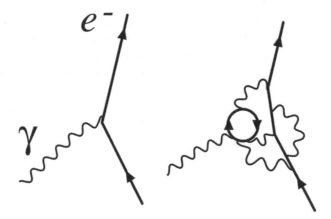

Figure 16.2 Renormalised vertex. Renormalisation in quantum electrodynamics: The simple electron/photon interaction that determines the electron's charge at one renormalisation point (left) is revealed to consist of more complicated interactions at another (right). https://en.wikipedia.org/wiki/Renormalization#/media/File:Renormalized-vertex.png.

Feynman's idea of a 'tower of effective field theories' ran up against the fundamental limit of spacetime. Feynman's diagrams, like Dirac's equations, assumed that particles moved through a *static* background spacetime. He assumed that time moved from left to right and space from bottom to top. So what was needed next was a quantum theory of spacetime.

One theoretical, but experimentally unproven, attempt to provide such a theory is loop quantum gravity. Loop quantum gravity uses a three-dimensional version of Feynman's diagram to imagine how quantum fields, not the particles in them, might interact to create three-dimensional space.[47] The ingenious result is shown in figure 16.3.

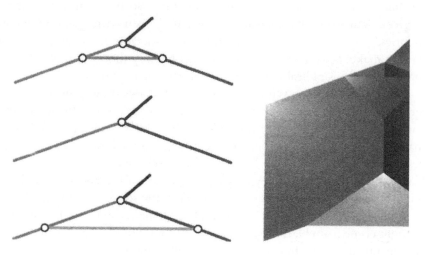

Figure 16.3 An evolving spinfoam network. From Carlo Rovelli, *Reality Is Not What It Seems* (New York: Penguin, 2017), 186.

As quantum fields vibrate and interact, they make space and time. In the equations of loop quantum gravity, one can model energies down to the lowest quantifiable limit of Planck's length (1.6×10^{-35} metres). This is significantly smaller than any known particle or possible experimental measurements. Below the Planck length, the fluctuations of energy become so radically indeterminate that if one tried to observe them, the energies of the photons required would be so powerful that they would create a black hole.

This is why Carlo Rovelli and other quantum gravity theorists describe the Planck length as a 'natural cut-off' to quantum theory.[48] Inside the quantum 'loops', 'bubbles' and 'foam', some theorists speculate, there is a vast and inaccessible 'ocean of microscopic black holes'.[49] For Rovelli, this justifies him in concluding that there is 'nothing'[50] below the Planck length and that space is made of discrete 'atoms' or 'grains' the size of Planckian units.[51] The Planck cut-off, for Rovelli, is not the same as renormalisation because it is not an arbitrary experimental length but a constant of nature.

Technically, though, something *does* happen below the Planck length. Energy does not vanish below the Planck length but instead becomes so radically indeterminate that the known laws of physics break down. For instance, if you put a particle in a box the size of the Planck length or smaller, the indeterminacy of its position would be greater than the size of the box, and its mass would produce a black hole whose radius would be double the Planck length. The time it would take to cross this radius would be four times the Planck time.[52] At these ultra-intense energies, the fluctuation and curvature of space becomes so indeterminate that we cannot calculate anything meaningful about it, even with quantum gravity theories.[53] Indeterminacy becomes more considerable than any prediction we can make. Statistical mechanics and probability theory run aground on the shores of this radical indeterminacy at the heart of things.

The kinetic operator is not an empty void but a seething generative indeterminacy. At the heart of black holes that dive below the Planck length, energy and momentum do not vanish but become so indeterminately high that even light cannot escape.[54] Black holes are not cosmic vacuums that destroy energy but weavers that unravel and reweave it.[55] If there are tiny Planck-sized black holes, they may be the source of the enormous indeterminate quantum fluctuations that make spacetime so foamy and unstable.[56]

One possible explanation for so much dark matter in the universe is that it might come from black hole radiation.[57] Black holes radiate both thermal and quantum energy. In particular, the smaller the size of a high-mass black hole, the larger its radiation of energy and the more dramatic the fluctuations of space and time created around it.[58]

The critical point of my interpretation is that there is a practical and theoretical limit to what quantum physics can meaningfully measure and objectify. That limit is indeterminate movement, the kinetic operator. My point is not that science has no access to these processes, but that these processes are part of measurement itself. Measurement is not something neutral that scientists do to the world, but something that the world does to itself. The kinetic operator resists objectification but is also active and immanent in all objects. It is the indeterminacy woven into objects, but which is irreducible to them.

Below the Planck length, quantum indeterminacy has *no higher explanation* because it exceeds any prediction method known to physics. Therefore, in my interpretation, we should not call indeterminacy 'random', 'deterministic' or 'probabilistic'. These are ideas imported from the modern theory of the object. My point is that indeterminacy might be pointing us in another direction. It is an entirely viable interpretation to suggest that the movement of matter/energy is not reducible to objects but is their immanent and non-totalisable condition. Science is always working at these limits to generate new objects, which can be a productive approach. However, indeterminacy has not been, and likely will never be, reducible to an object.

This does not mean that the study of objects and the prediction of their patterns is pointless. On the contrary, it can be generative and practical. Science does not need to explain indeterminacy by something determinate to get useful results. There is another way. Science can accept the generative indeterminacy of matter as the immanent origin of determinacy. It can affirm that indeterminacy actively *resists and assists* in the creation of objects as their *kinetic operator.* This would be a significant interpretive break with prior theories of the object, but it would still be consistent with contemporary physics.

We can even reinterpret the core features of quantum theory in this light using the kinetic vocabulary proposed in this book.

Flows

Earlier, I defined a flow as an ongoing movement or process without substance, discreteness or fixed properties. This is what quantum theory studies. As Rovelli says, 'In the world of quantum mechanics, everything vibrates; nothing stays still. The impossibility of anything being entirely and continuously still in a place is at the heart of quantum mechanics.'[59] The quantum world is one of processes and patterns, and one cannot study it without entering into the flow. The primacy and ubiquity of movement in quantum theory is a theoretical insight of great significance for the loop theory of objects.

Movement is not only primary in the quantum world; it is also, as we discussed above, *globally entangled.* It is not possible to *fully* account for movement in terms of local measurements[60] because every local measurement is always correlated with non-local flows.[61] Non-local indeterminism is the idea that the entire universe flows together. Instead of explaining this entanglement by global determinism or treating it as random, we can take an anti-essentialist perspective and acknowledge that entanglement and indeterminism are irreducible phenomena.[62]

If motion is primary and entangled, then all localities change together *simultaneously* as regions of the whole flowing universe without cause.[63] Entanglement does not mean that there is any 'faster-than-light communication'[64] between localities. Its patterns of motion are not caused by something else outside the universe, but instead, the universe is the entangled cause of itself.

In this interpretation, the universe flows as an open and dynamic whole. It is not a deterministic wave equation given in advance or just one of many possible worlds (see figure 16.4). One consequence of kinetic indeterminacy is that there is no probability set large enough to include the universe's flow. At every level of nature, we deal with indeterminate flows that we cannot fully understand as local, discrete or totalisable.[65] As Karen Barad writes,

> Matter's dynamism is generative not merely in the sense of bringing new things into the world but in the sense of bringing forth new worlds, of engaging in an ongoing reconfiguring of the world. Bodies do not simply take their places in the world. They are not simply situated in, or located in, particular environments. Rather, 'environments' and 'bodies' are intra-actively co-constituted. Bodies ('human', 'environmental', or otherwise) are integral 'parts' of, or dynamic reconfigurings of, what is.[66]

Any measurement of the universe is entangled with what one is measuring – this is why there are no neutral observers but only kinetic operators constitutive of scientific objects.[67]

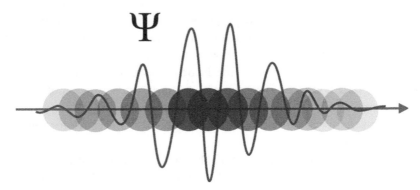

Figure 16.4 Quantum wave equation. The quantum wave equation can be represented as a superposition of possible positions a particle may be in. The positions in the centre of the wave diagram occur with higher frequency than those on the left or right. The wave equation is represented with the Greek letter Ψ. https://simple.wikipedia.org/wiki/Interpretations_of_quantum_mechanics#/media/File:Qm_template_pic_4.svg.

Folds

The kinetic operator also transforms how we interpret what a *measured* quantum object is. Measurement does not discover pre-existing objects but participates in making them by folding indeterminate fields together.

Physicists' actions and experimental apparatuses do not discover discrete objects but co-produce metastable measurement-events.

If matter is globally entangled, then all measurements and objects are also non-local acts. This means that quantum objects are more like events of spatiotemporal metastability or folds of spacetime.

For instance, in 'entropic gravity theory', objects are understood as regions of entangled and enfolded movement. The higher the degree of entanglement in a particular region in the quantum vacuum, the higher the region's entropy. This entropy 'turns out to be naturally proportional to the area of the region's boundary. The reason isn't hard to understand: field vibrations in one part of space are entangled with regions all over, but most of the entanglement is concentrated on nearby regions.'[68] In other words, the more entanglement there is, the higher the entropy, and the closer the fluctuations are to one another. The weaker the entanglement, the more distant. According to this theory, space and the laws of general relativity come from entropic and entangled motion.[69] Furthermore, energy, entropy and entanglement do not *go away* below Planck's constant. They just become increasingly indeterminate. As they become less entangled, less entropic and less indeterminate, they become more determined, metastable and relatively discrete. Objects are born from this kinetic operation of indeterminate folding and not from smaller fundamental units.

Fields

How big is the range of indeterminate movements for a given event? Where are the limits of the event? In quantum physics, there is no fixed or absolute answer to these questions. If matter is genuinely indeterminate and globally entangled, we cannot draw any firm lines at the beginning and end of an event. When we draw these lines and choose to select some portion of reality to study, we bend and shape our object of study.

Kinetically, quantum physics does not explain the entanglement and indeterminacy of matter. Instead, it *works with it*, cutting, bending and shaping it into relatively stable regions called ranges of possibility. These ranges are what I call kinetic fields. The fields do not explain motion but perform and weave it. As the American philosopher Arthur Fine argues, the demand that we should explain motion by something else

represents an explanatory ideal rooted outside the quantum theory, one learned and taught in the context of a different kind of physical thinking. It is like the ideal that was passed on in the dynamical tradition from Aristotle to Newton, *that motion as such requires explanation*. As in the passing of that ideal, we can learn from successful practice that progress in physical thinking may occur precisely when we give up the demand for explanation, and shift to a new conception of the natural order.[70]

In my view, the way towards a new kinetic theory of the object is to abandon our efforts to *explain* indeterminate entangled motion by determinism, randomness or probability, and instead to reinterpret nature as fundamentally kinetic. Quantum science works perfectly well by tracing the patterns of matter in motion without seeking after their ultimate explanation by something else. This is what I mean by interpreting quantum physics as a 'kinetic field'. It is an immanent field or range that maps and co-creates the movement of matter without explaining it by some more profound principle or fundamental measurement unit.

Accordingly, I disagree with interpreting the kinetic field as a series of linear, superpositioned, deterministic sequences where one series is actualised when the wave function 'collapses' probabilistically. This view assumes that the movement of matter merely seems indeterministic but is really deterministic.[71] In this way, some scientists want to explain local indeterminacy by non-local deterministic processes.

The kinetic field is neither random, deterministic nor probabilistic. It is a Brownian 'walk' in which each movement is related to, but not determined by, what came before. A distribution of objects is an open and continually changing range of motions. Interpreting the distribution of objects in this way might allow scientists to more clearly see the generative and indeterminate relations that can emerge in fields if they do not try to explain them by some fixed set of possibilities.

What happens to a particle between two measurements? One way to interpret what happens is to give the complete set of all possible paths that the particle could have travelled. In this view, each track is entirely deterministic, but the particle only follows one random route from the complete set of paths. I propose a different interpretation. What happens between each measurement is indeterminate down to the Planck length and below. There is no need to try and explain away this indeterminacy with an image of every deterministic path that might have happened between measurements.

Instead, we can call this unmeasured region the kinetic operator, mark it with a ∇, and move on without trying to explain it away. We could also make more measurements between the first and second to refine the path. But the critical point is that there will always be a region between two measures that will be unmeasured. I propose that instead of interpreting that gap as a superposition of possible paths, we acknowledge it *as indeterminate*. Furthermore, we can admit that our measurements are not opposed to this indeterminacy but are metastable folds *of this indeterminate field*.

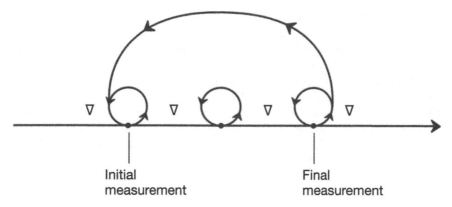

Figure 16.5 Quantum kinetics. Author's drawing.

We could call this kinetic interpretation an 'indeterminate realism'. The movement of matter is 'irreducibly'[72] indeterminate and entangled in co-constituting quantum walks.[73] Quantum fields *really* are continually and collectively modulating, folding and unfolding, and the kinetic operator *really* does participate in this collective determination via the measurement process.

The next chapter looks at how we might also interpret the mathematical-logical science of category theory from a kinetic perspective.

Notes

1. Including Richard Feynman and Julian Schwinger.
2. Lee Smolin, *The Trouble with Physics: The Rise of String Theory, the Fall of a Science, and What Comes Next* (London: Penguin, 2008), 62.
3. Carlo Rovelli, *Reality Is Not What It Seems: The Elementary Structure of Things*, trans. Simon Carnell and Erica Segre (New York: Penguin, 2017), 128.
4. John Polkinghorne, *Quantum Theory: A Very Short Introduction* (Oxford: Oxford University Press, 2002), 74.
5. Smolin, *The Trouble with Physics*, 128.
6. Frank Close, *The Infinity Puzzle: The Personalities, Politics, and Extraordinary Science Behind the Higgs Boson* (Oxford: Oxford University Press, 2013), 23.
7. Sean Carroll, *The Big Picture: On the Origins of Life, Meaning, and the Universe Itself* (New York: Dutton, 2017), 435.
8. See Rovelli, *Reality Is Not What It Seems*, ch. 9 for a full account of possible empirical confirmations.
9. Smolin, *The Trouble with Physics*, 146.
10. Polkinghorne, *Quantum Theory*, 74.
11. Close, *The Infinity Puzzle*, 5.

12. Stephen Battersby, 'It's Confirmed: Matter is Merely Vacuum Fluctuations', *New Scientist*, 20 November 2008, available at <https://www.newscientist.com/article/dn16095-its-confirmed-matter-is-merely-vacuum-fluctuations/> (last accessed 4 March 2021).
13. Rovelli, *Reality Is Not What It Seems*, 134.
14. Smolin, *The Trouble with Physics*, 55.
15. On the issue of local hidden variables, see David Mermin, 'Is the Moon There When Nobody Looks? Reality and the Quantum Theory', *Physics Today*, 38.4 (1985), 38–47; and R. E. Kastner, Stuart Kauffman and Michael Epperson, 'Taking Heisenberg's Potentia Seriously', *International Journal of Quantum Foundations*, forthcoming, available at <https://arxiv.org/abs/1709.03595> (last accessed 4 March 2021).
16. Polkinghorne, *Quantum Theory*, 73.
17. Rovelli, *Reality Is Not What It Seems*, 124.
18. Close, *The Infinity Puzzle*, 42.
19. 'To be entangled is not simply to be intertwined with another, as in the joining of separate entities, but to lack an independent, self-contained existence. Existence is not an individual affair. Individuals do not pre-exist their interactions; rather, individuals emerge through and as part of their entangled intra-relating.' Karen Barad, *Meeting the Universe Halfway: Quantum Physics and the Entanglement of Matter and Meaning* (Durham, NC: Duke University Press, 2007), ix.
20. Polkinghorne, *Quantum Theory*, 74.
21. '[Matter] has no property in itself, apart from those that are unchanging, such as mass. Its position and velocity, its angular momentum and its electrical potential only acquire reality when it collides-interacts with another object. It is not just its position that is undefined, as Heisenberg had recognized: no variable of the object is defined between one interaction and the next. The relational aspect of the theory becomes universal.' Rovelli, *Reality Is Not What It Seems*, 122.
22. See Erwin Schrödinger, 'An Undulatory Theory of the Mechanics of Atoms and Molecules', *Phys. Rev.*, 28.6 (1926), 1049: 'The wave-function physically means and *determines* a continuous distribution of electricity in space.'
23. Rovelli, *Reality Is Not What It Seems*, 122.
24. Close, *The Infinity Puzzle*, 43.
25. Rovelli, *Reality Is Not What It Seems*, 187.
26. 'The closed lines that appear in the solutions of the Wheeler-DeWitt equation are Faraday lines of the gravitational field.' Rovelli, *Reality Is Not What It Seems*, 161.
27. Rovelli, *Reality Is Not What It Seems*, 187.
28. Rovelli, *Reality Is Not What It Seems*, 135.
29. Barad, *Meeting the Universe Halfway*, 14.
30. 'Whenever you have a diagram where lines form one or more closed loops, the infinity disease erupts.' Close, *The Infinity Puzzle*, 47–8.
31. There is an extensive literature on this phenomenon. See Mermin, 'Is the Moon There When Nobody Looks?'; Amir Aczel, *Entanglement: The Unlikely*

Story of How Scientists, Mathematicians, and Philosophers Proved Einstein's Spooki-est Theory (New York: Plume, 2003); Louisa Gilder, *The Age of Entanglement: When Quantum Physics Was Reborn* (New York: Alfred A. Knopf, 2009).

32. E. Schrödinger, *Naturwissenschaften*, 23, 844 (1935); English translation in J. A. Wheeler and W. H. Zurek (eds), *Quantum Theory and Measurement* (Princeton: Princeton University Press, 1983), 152.

33. Carroll, *The Big Picture*, 100.

34. Albert Einstein, Hedwig Born and Max Born, *The Born–Einstein Letters: Correspondence between Albert Einstein and Max and Hedwig Born from 1916–1955, with Commentaries by Max Born* (New York: Walker, 1971), March 1948.

35. David Bohm's earlier non-local hidden-variable theory was inspired by de Broglie's heterodox pilot-wave theory from the 1920s.

36. Mermin, 'Is the Moon There When Nobody Looks?'

37. Mermin, 'Is the Moon There When Nobody Looks?' See also Peter Holland, *The Quantum Theory of Motion: An Account of the Broglie-Bohm Causal Interpretation of Quantum Mechanics* (Cambridge: Cambridge University Press, 2004).

38. Holland, *The Quantum Theory of Motion*.

39. For a similar, but distinct, proposal, see Arthur Fine, 'Do Correlations Need to be Explained?', in J. Cushing and E. McMullin (eds), *Philosophical Consequences of Quantum Theory* (Notre Dame: University of Notre Dame Press, 1989), 175–94; Tanya Bub and Jeffrey Bub, *Totally Random: Why Nobody Understands Quantum Mechanics (A Serious Comic on Entanglement)* (Princeton: Princeton University Press, 2018); B. van Fraassen, 'The Charybdis of Realism: Epistemological Implications of Bell's Inequality', *Synthese*, 52 (1982), 25–38.

40. Henri Bergson, *Matter and Memory* (New York: Zone Books, 2005), 104.

41. Rovelli, *Reality Is Not What It Seems*, 136.

42. Close, *The Infinity Puzzle*, 42.

43. The solution, in the case of quantum electrodynamics, by S.-I. Tomonaga, J. Schwinger and R. Feynman, and others, involved far more than just relying on such measurement. The renormalisation of the electroweak theory (which unifies QED and the theory of weak interactions) was done by M. Veltman and G. t'Hooft.

44. Close, *The Infinity Puzzle*, 8.

45. Jeremy Butterfield and Nazim Bouatta, 'Renormalization for Philosophers', *Poznan Studies in the Philosophy of the Sciences and the Humanities*, 104 (2016): 437–85.

46. Close, *The Infinity Puzzle*, 60.

47. Rovelli, *Reality Is Not What It Seems*, 189.

48. '[F]unction in the theory as a natural IR cut-off, next to the UV cut-off provided by the Planck length.' Carlo Rovelli, *Covariant Loop Quantum Gravity: An Elementary Introduction to Quantum Gravity and Spinfoam Theory* (Cambridge: Cambridge University Press 2020), 118.

49. Stephen Hawking, 'Virtual Black Holes', *Physics Review*, D 53 (1996), 3099; G. t'Hooft, 'Virtual Black Holes and Space-time Structure', *Foundations of Physics*, 48 (2018), 1134–49; Rovelli, *Reality Is Not What It Seems*, 248.

50. 'Below [the Planck scale], nothing more is accessible. More precisely, nothing exists there.' Rovelli, *Reality Is Not What It Seems*, 152.

51. For a discussion of these values or 'eigenvalues' and how to derive them, see Rovelli, *Reality Is Not What It Seems*, 165.

52. Ethan Siegel, 'What is the Smallest Possible Distance in the Universe?', *Forbes Magazine*, 26 June 2019, available at <https://www.forbes.com/sites/startswithabang/2019/06/26/what-is-the-smallest-possible-distance-in-the-universe/#7fdb98c548a1> (last accessed 4 March 2021).

53. This is what Susskind calls 'black hole complementarity'. Leonard Susskind, *The Black Hole War: My Battle with Stephen Hawking to Make the World Safe for Quantum Mechanics* (New York: Back Bay Books, 2009), 237.

54. See Christopher N. Gamble and Thomas Nail, 'Blackhole Materialism', *Rhizomes: Cultural Studies in Emerging Knowledge*, forthcoming.

55. Stephen Hawking, *A Brief History of Time* (New York: Bantam Books, 2017); Rovelli, *Reality Is Not What It Seems*, 248; Susskind, *The Black Hole War*, 124–5.

56. Susskind, *The Black Hole War*, 98–9. See t'Hooft, 'Virtual Black Holes and Space-time Structure.'

57. 'Unless dark matter is composed of black holes, which is an open possibility.' Rovelli, *Reality Is Not What It Seems*, 128.

58. Susskind, *The Black Hole War*.

59. Rovelli, *Reality Is Not What It Seems*, 226.

60. Abner Shimony, 'Bell's Theorem', *The Stanford Encyclopedia of Philosophy* (Fall 2017 edition), ed. Edward N. Zalta, available at <https://plato.stanford.edu/archives/fall2017/entries/bell-theorem/> (last accessed 4 March 2021).

61. Rovelli claims to have sidestepped the problem of entanglement because the act of measurement is always local. Furthermore, for Rovelli, at the Planck scale spacetimes are separated by immeasurable black holes, and thus ultimately discontinuous from one another. See Carlo Rovelli, 'Loop Quantum Gravity', in Bernard D'Espagnat and Hervé Zwirn (eds), *The Quantum World: Philosophical Debates on Quantum Physics* (Cham: Springer, 2017), 279–94.

62. See Fine, 'Do Correlations Need to be Explained?' However, I disagree with Fine's conclusion that indeterminacy is random. I thank Christopher N. Gamble for his critique of Fine on this point.

63. 'Localization assumes for its formulation the existence of separate states, but they need not be of the kind assumed by the separability principle; that is to say, they need not be such as to determine completely the joint state of every composite system to which the systems they characterize may belong as parts. Thus, it is possible to have a local, but nonseparable theory, quantum mechanics being the most important example.' Don Howard, 'Holism, Separability and the Metaphysical Implications of the Bell Experiments', in J. Cushing and E. McMullin (eds), *Philosophical Consequences of Bell's Theorem* (Notre Dame: University of Notre Dame Press, 1989), 224–53 (277).

64. See J. Henson, 'How Causal Is Quantum Mechanics?', talk given at New Directions in Physics Conference, Washington, DC, 2015, available at

<http://carnap.umd.edu/philphysics/hensonslides.pptx> (last accessed 4 March 2021).

65. Thanks to Christopher N. Gamble for pointing out that this is an unstated but direct and necessary implication of Karen Barad's philosophy: that all random/probabilistic/deterministic accounts of matter render matter inherently passive and non-creative. See also Christopher N. Gamble, Joshua S. Hanan and Thomas Nail, 'What is New Materialsm?', *Angelaki*, 24.6 (2019), 111–34 (113–15).

66. Barad, *Meeting the Universe Halfway*, 63.

67. The 'measurement problem' has been treated in detail elsewhere, so I will not repeat the standard story here but only refer the reader to the existing literature.

68. See Sean Carroll, *Something Deeply Hidden: Quantum Worlds and the Emergence of Spacetime* (Boston: Dutton, 2019).

69. For more on a review of entropic gravity and its relationship to loop quantum gravity, see Lee Smolin, 'Newtonian Gravity in Loop Quantum Gravity', *Perimeter Institute for Theoretical Physics*, 29 October 2018, available at <https://arxiv.org/pdf/1001.3668.pdf> (last accessed 4 March 2021).

70. See Fine, 'Do Correlations Need to be Explained?', 192, my italics.

71. Others reject the 'collapse' of the wave function in favour of deterministic worlds that 'branch' into parallel universes instead of 'collapsing'. See Carroll, *Something Deeply Hidden*.

72. Fine, 'Do Correlations Need to be Explained?', 193.

73. 'The probabilistic nature of the result is rooted in ontological indeterminacy and not classical ignorance.' Barad, *Meeting the Universe Halfway*, 268.

17

The Contemporary Object II: Category Theory

The second contribution to a loop theory of the object emerged from the mathematics of category theory in the 1940s. Category theory was a direct response to set theory.[1] The idea of a 'category' is not just a replacement or refinement of the idea of a 'set'. In my interpretation, it is a profound conceptual change to the modern theory of the mathematical object. In this chapter, I want to show how category theory directs us to a new kinetic and indeterministic theory of objects.

Unlike set theory, category theory does not attempt to ground mathematical activity in an ontology of types of objects such as sets. It is a logic of the continual transformations *between mathematical objects*. Early moves towards such a mathematical theory of 'relations of transformation' came from at least two sources. On the one hand, category theory 'began as a project to study continuous mappings within the program of algebraic topology'.[2] This effort was motivated by Poincaré's attempts to study continuous differential equations qualitatively and Riemann's theory of manifolds, discussed in Chapter 14.[3]

On the other hand, category theory was also a response to a historical crisis in the set-theoretical foundations of mathematics. In the 1920s, versions of the Lowenheim–Skolem theorem argued that if the axioms of set theory were a 'countable set of axioms', and countability was an absolute notion, then set theory could only prove a countable model of mathematics. It would, therefore, not be able to prove the existence of a model with *uncountable* sets, as set theory wants to. There have been many different attempts to resolve and dissolve the paradox, but one consequence stuck with mathematicians. There is no absolute notion of countability that proves both countable and uncountable sets.[4]

In 1931 Gödel's incompleteness theorem further shattered set theory's pretensions as an ultimate foundation for mathematics. He showed that for any formal system powerful enough to model arithmetic, there will always be statements that are true concerning its axioms, but that can neither be proved nor disproved within the system.[5] More radically, any 'effective' or formally recursive set of axioms cannot prove, by using only its own axioms, that there cannot be at least one more axiom that could be added to form a new consistent system. Therefore, axiomatic set theory (Zermelo–Fraenkel set theory) had increasingly to give up its hopes of consistency, universality and foundationalism.

These problems gave rise to an alternative tradition in the history of mathematics called constructivism. Constructivism is an anti-Platonic and anti-formalist theory of mathematics based on the practice of constructing internally consistent systems. According to this tradition, mathematics is not the discovery of fundamental principles but rather the creation of new structures. The existence of mathematical objects must be done and shown *in practice* and not merely assumed true by the law of non-contradiction or by *reductio*.

Historically, mathematical constructivism began with the early work of 'intuitionists' or 'subjective constructivists', including the Dutch mathematicians Jan Brouwer and his student Arend Heyting in the 1920s–1940s. Over the twentieth century, constructivism also gave rise to significant new mathematical research programmes beyond early subjective interpretations.[6] The best-known of these today is category theory.

The American mathematicians Samuel Eilenberg and Saunders Mac Lane co-founded category theory in the 1940s. Although they were not explicitly constructivists, category theory has several constructivist features.[7] This is because category theory, like constructivism, was a continuation of a long legacy that began as a reaction against Dedekind and Cantor's increasingly abstract Platonism and Hilbert's formalism. Constructivism was a response to the recognised impossibility of absolute completeness, absolute countability and foundationalism in mathematics in the twentieth century. In category theory, for example, the axiom of choice is not always valid, and the law of excluded middle does not always hold. The law of excluded middle says that for any proposition, either that proposition is true or its negation is true. There is no third or middle option.

Category theory's constructivism is not the result of any direct influence from constructivists. It results from taking Gödel's incompleteness theorem seriously and trying to develop a new meta-mathematical system with it in mind. Such an approach ultimately has to give up on foundational formalisms and become de facto constructivist.

This chapter aims to offer a kinetic interpretation of three core components of category theory: its hybridity, pedesis and relationalism. These are also three aspects of the loop object I want to outline.

Hybridity

Category theory is not a mathematical foundation as much as it is a mathematical system of hybridity. It is broad enough to formalise all previous mathematical systems but also open enough to allow for the creation of new systems that hold an increasing number of radically different axioms. In category theory, for instance, there is no absolute a priori foundation for mathematics but rather different categories or mathematical regions with their own set of coherent rules. Set theory is only one of these regions. More on this below.

The key point I want to highlight in my interpretation is that the foundation of any mathematics has no existence independent of its *immanent kinetic construction*. Category theory is so hybrid, in part, because it gives up on trying to be a foundation for formal objects 'in themselves'.

Category theory is such a fascinating and significant theory of mathematical objects because of its emphasis on the *structural relations* or *movements* immanent to and constitutive of mathematical objects. It is a theory of objects that begins not with the pre-existence of a foundational type of object (ordinal, cardinal, intensive or potential), but with the constructive action of *morphism* itself.

As the French philosopher and mathematician Alain Badiou writes,

> To show in any given universe that a particular object 'exists', we must somehow show it (that is, 'construct' it, as always, from the arrows of which it is the source and the target); it will not suffice to establish that its non-existence is contradictory. There is therefore in this kind of thought no indirect (or oblique) proof of existence or truth. We are only bound to believe what we see (or construct).[8]

'Category theory', Badiou continues, 'is entirely relativist, it shows a plurality of possible universes.'[9]

The Colombian mathematician Fernando Zalamea describes this plurality as a 'mixed mathematics' of 'mixed constructions' that 'links theories and ideas between different subdomains of mathematics, with the consequent transformation of objects when they are read from multiple variable contexts'.[10] This is possible because category theory treats all mathematical systems as purely structural and relational networks, without reducing them to a single type of object. It is '[t]he legitimacy of a theory of abstract structures, independent of the objects connected together by these structures', as the French philosopher of mathematics Albert Lautman would say.[11] What an object 'is' changes with respect to the category and relations in which it is incarnated.[12] This is the source of what Badiou calls its 'universal power' in contemporary mathematics.[13]

What is a category? A category is a mathematical structure or grouping of 'objects' connected by relations called 'arrows' or 'morphisms'. In category theory, anything can be an object (numbers, letters, places and names), and one can describe any relation by using arrows between them (space, time and motion). A category is a general structure shared by objects and their relations. For example, the category of sets has sets as its objects and functions as its relations. A function maps elements from Set A to elements in Set B. There is also a category of groups, with groups as objects and homomorphisms as relations, and a topology category with topological spaces as objects and continuous functions as relations.

One can also coordinate different categories with one another in relations called *functors*. Functors are the relations between categories that establish one-to-one correspondences between the objects and relations between different categories. Therefore, functors are essential to the meta- or hybrid nature of category theory. More on this below.

The diagram

Category theory is also hybrid because of the importance of its 'diagrams'. Diagrams are regions or pieces of a category that have objects and arrows. For mathematical objects to exist in category theory, one has to *show* or *describe* them via diagrams. Objects have *no a priori existence* or properties. In this way, the diagrams drawn by mathematicians do not assume a fundamental division between presentation and representation, as set theory does. In set theory, there is a fundamental disjunction or excess of counting over what is counted. A set is supposed to be defined by its elements and can always be embedded in a larger set. However, Bertrand Russell and the Italian mathematician Cesare Burali-Forti showed that 'the second in fact *exceeds* the first in an errant or immeasurable way'.[14] No set includes itself among its elements.

Category theory's diagrams are different because they do not assume any a priori existence of objects that one represents. The reality of objects is immanent with their description: 'In the categorical formulation we find *neither* the foundational dimensions of a relation of belonging, nor the disjunction between two modalities of being-in.'[15] In category theory, 'Diagrams thus represent systems of relations and at the same time instantiate (at least some of) those relations directly. In this way the "content" of a diagram is already at least partly present directly and immediately in its "form".'[16]

Another consequence of category theory's constructivism is that one cannot determine an object's existence or the truth of a proposition in advance of a diagram showing the relations under which these would be the case. This leads to, among other things, a violation of the law of excluded middle. Assuming the 'law of excluded middle', we typically think that for any proposition, either that proposition is true or its negation is true. There is no third position. However, in category theory, just because we negate something does not necessarily prove that its opposite is true. We cannot deduce proof formally or logically by negation alone. It requires a positive demonstration with a diagram. For example, the disjunctive claim that *either $x = 0$ or $x \neq 0$* is not *necessarily true*. It is only true if one shows it with a particular diagram. This constructivism also entails a fundamental shift in the theory of identity, which we will explore in the following sections.[17] Category theory is hybrid because its diagrams are not closed by any other fundamental constraints such as the law of excluded middle or the axiom of choice. Categories are hybrid because they are open and without an absolute limit object.

Pedesis

Category theory is an especially kinetic mathematics because, unlike formalism, the existence of mathematical objects, in my definition (including categories, functors, arrows and categorical objects), is ontologically *indeterminate* until constructed in a concrete singular act. Nothing is or can be assumed until one enacts it. This sounds odd because the Western tradition has understood mathematics as formalism for so long. Mathematics has often sought foundations in the law of identity, non-contradiction, the law of excluded middle and the axiom of choice.

Category theory is different because, at least in my interpretation, it takes seriously the kinetics of coordination, which is *constitutive* of mathematical knowledge. Mathematics is an embodied *and* constructive activity that flesh and blood mathematicians have to *do*. Knowledge is doing and requires movement. Truth is immanent to its construction.

Process mathematics

Category theory is a 'process mathematics' without foundations. As Alain Badiou writes,

> Category theory determines being as act (or relation, or *movement*). Its basic concept is the arrow (or morphism, or function) which 'goes' from one object to another object. We should not be misled

by this objectual vocabulary: an 'object', in category theory, is in the first place a simple point (or a simple letter) without any determinate interior, while a set is precisely nothing other than the count-for-one of what belongs to it, namely, its 'interior'.[18]

The objects of category theory have no essence or 'interiority' because their existence is immanent with their construction by arrows or movements.

The most basic category is the 'empty category', which contains no objects or arrows. The primary agent in categories is the 'arrow', which is also called a morphism or continuous mapping. Mathematicians draw it as a continuous line with a directional arrowhead. In contrast to the modernist elasticity of the discrete set theoretical object which 'subordinates movement (the functions) to a fixed position of presentation', the category object is, as Badiou writes, 'identified only through its movements, through the operations performed, movements and operations that involve the entire universe (all the other "objects", but above all the "forces" or arrows which criss-cross this universe)'.[19]

These continual movements (morphisms) are pedetic insofar as they are not defined in advance but rather successively or additively. It is always possible to add one more axiom, arrow or object to a category. This is radically unlike an arithmetical series or most sets, because the previous object does not determine each new arrow. Each new arrow can change the nature of the earlier objects. 'The categorial procedure is itself descriptive. It proceeds by successive definitions, which enable the specification of possible universes.'[20] In this respect, category theory follows in the footsteps of a pedetic constructivism. Future arrows are not entirely determined by the history of what came before but are always related from somewhere (the source) and headed towards somewhere (the target). Morphisms are thus neither random, determined nor probabilistic, but pedetic or 'indeterminate'.

Arrows are more primary than objects. Eilenberg and Mac Lane show that category theory works equally well with 'arrows only'.[21] One can remove the object entirely, and the movements (arrows) retain the structure of relations that define the objects. The practical primacy of these pedetic arrows introduces what Zalamea calls an 'ontological fluctuation' or 'vast liquid surface'[22] deeply concerning to 'certain analytic approaches to the philosophy of mathematics, which seek to delimit and pinpoint their perspectives in the clearest possible way, fleeing from smears and ambiguities, and situating those tidy delimitations over fragments of the absolute'.[23]

The performative nature of category theory introduces 'mobile frontiers between the conceptual and the material'.[24] The mathematical object becomes one with the act of mathematical creation itself. The French philosopher of mathematics Jean Cavaillès describes it as material and kinetic 'gesture'.[25] As Zalamea says,

> Objects in this realm cease to be fixed, stable, classical and well-founded – in sum, they cease to be 'ones'. Instead they tend toward the mobile, the unstable, the nonclassical, and the merely contextually founded – in short, they approach 'the many.' Multiplicity everywhere underlies contemporary transit and the objects of mathematics.[26]

Objects in category theory become 'webs and processes. Determinate "entities" firmly situated in one absolute, hard-and-fast universe, do not exist; instead, we have complex signic webs interlaced with one another in various relative, plastic and fluid universes.'[27]

Category theory is therefore a strange kind of mathematics in which the foundation has become an 'indeterminate' and 'mobile base'.[28] This is precisely why the American physicist Lee Smolin writes in *Three Roads to Quantum Gravity* that category theory is 'the right form of logic for cosmology'.[29] He even says that category theorists beat quantum field theorists to the mathematics of the quantum world. Category theory is the 'logic for the working cosmologist', Smolin writes, playing on Mac Lane's book title *Categories for the Working Mathematician* (1971). It can deal with indeterminacy and intra-active 'observer dependence' in ways that classical formalist mathematics cannot. In category theory, 'Statements can be not only true or false; they can also carry labels such as "we can't tell now whether it's true, but we might be able to in the future"'.[30] The performance of mathematics, just like a quantum observation performance, is immanent to and constructive of the real existence of mathematical entities.

Feedback

Categorical objects are also profoundly relational. They are not pre-existing objects that relate to one another, but instead arrows that have folded or looped back over themselves. The constructivist nature of category theory requires that mathematical objects be immanent (i.e. not transcendent) to their real, practical and performative description by the mathematician. When an arrow folds back over itself, it folds over itself in the moment of a practical inscription in the diagram. Furthermore, each time one adds a new relation, the previous objects change in the diagram. They become different objects responsive to real-time feedback in the practice of mathematical creation.

Loops

A mathematical object is a self-interaction. The diagram of the interaction shows the metastable 'identity' of the categorical object.

This is why categorical objects only play a *secondary* role for Eilenberg and Mac Lane. Its object is always a loop-object or 'quasi-object',[31] defined entirely by the immanent arrow that produces its identity. The secondary character of identity is therefore 'inevitable', Badiou says, 'since identity is not an immanent mark, but rather the effect of actions, or arrows, which operate in a categorial universe'.[32] This strange form of self-identity in the categorical object introduces a tension into mathematics between stability and morphism.

The American category theorist William Lawvere writes that the theory of categories was 'the first to capture in reproducible form an incessant contradiction in mathematical practice: we must, more than in any other science, hold a given object quite precisely in order to construct, calculate, and deduce; yet we must also constantly transform it into other objects'.[33] This is achieved by treating an object as immanent to the movement or morphism of its own identity arrow and allowing this identity to be continuously changed with respect to its ongoing diagram with others. The identity of the object is immanent to its extrinsic relations, including its own extrinsic self-relation loop.

In other words, as Badiou says, 'in a categorial universe an object is determined exclusively by the relations, or *movements*, of which this object is the source or the target'.[34] An object emerges when an arrow becomes the target and source of itself – a kind of iterative or recursive self-function.[35]

In category theory, the object's identity is not a fixed essence because the object is only defined 'up to isomorphism'. This means that many different objects or arrows can all operate as the 'same' object. For example, an object can be a 'monoid' in which multiple arrows have the same object as their target *and* source. Each arrow is different and defines another *aspect* or feature of 'the same' object.

Webs

Categorical objects are not static objects but regions or nodes defined by their performative relations. Objects are not fixed into these relations but rather change as the web of relations changes. A categorical object, as Zalamea writes, 'is not something that "is", but something that is in the process of being'.[36] For example, in category theory, if b is a morphism (f) of a, then a is defined relationally as the source of f and b as the target. Both a and b are folds in a continuous relation of f. However, if we add 0 as the

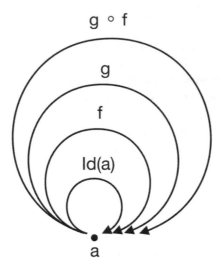

Figure 17.1 Category theory monoid. Author's drawing.

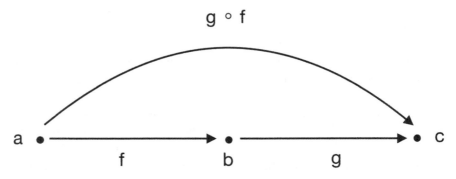

Figure 17.2 Category theory composition. Author's drawing.

source of a new morphism, *g*, and make *a* its target, the whole relation changes to one of composition.

All three objects change with the introduction of the new relation of *0(g)*. Our first term *a* is transformed from being only a source object to a source and a target object. Our second object, *b*, is transformed from being merely the monomorphic target of *a* to being the dual-morphic target of both *a(f)* and *0(h)* such that:

$$h \circ (f \circ g) = (h \circ f) \circ g$$

Our third object, *0*, becomes a new central or initial object or universal perspective on the whole diagram, of which *b* is the 'limit', 'terminal object' or 'universal property' of the diagram.

Thus, category theory objects emerge immanently to their constitutive and changing non-local relations, operations, actions or movements. Category theory attempts to describe objects according to the 'world's continuous becoming and the continuous enfolding [as a] relational canopy', or 'ambient milieu' of 'tactical and functional webs of coupling'.[37]

Kinos theory

Category theory makes several radical moves away from the modernist theory of the object dominated by set theory. It offers a new model of hybridity without foundationalism. Instead of formalist universality and determinism, it offers a pedetic and constructivist model capable of dealing with 'indeterminate' or 'vague' 'quasi-objects'. I define these below. And instead of isolated static objects, it introduces relational objects defined by ongoing feedback immanent to the practice of mathematical description itself.

In this section, I want to interpret category theory through the lens of the kinetic theory of the object. I use the concepts of flow, fold and field to synthesise the general operations of how objects are created in category mathematics. I also use the idea of a 'kinetic operator' to highlight the indeterminate nature of motion that I think leads towards a new theory of objects.

In recent versions of category theory, some mathematicians are motivated by the impossibility of describing the indeterminate objects discovered in quantum physics. Some of these mathematicians have developed the idea of 'quasi-sets', in set theory, and 'quasi-categories', in category theory. Quasi-objects do not assume that the principle of identity is universal.[38] Quasi-objects are neither self-identical nor necessarily equal to anything else in logic. They are understood strictly indeterminately and relationally.

I want to look at three places in category theory where this indeterminate kinetic operator is at work. In what follows, I want to show how the mathematical domains of categories, sets and topoi are all subregions of what category theorists call a 'weak higher category' and what I am calling kinos theory. A weak higher category means that any composition of arrows need not be uniquely defined but can be composed of an infinity of other compositions that cannot be well defined. Kinos theory is my term for what the Canadian category theorist André Joyal calls a 'meta-stable quasi-category'.[39] A metastable quasi-category is one where every object in it is 'infinitely connected' and 'infinitely composed'. I will say more about this below. I am using the term 'kinos' here to emphasise the importance of movement and metastability in the objects of higher category theory.

Flow

The first site of indeterminacy in category theory is in its arrows. Category theory treats arrows as both mathematically primary and existentially performative. By this, I mean that they must be drawn by someone to exist. Classical category theory also treats arrows and their compositions as unique. But what happens if the arrows are indeterminate and non-unique?

This is a strange question because arrows, in category theory, are supposed to be the relations or processes of morphism that make *identity* possible by looping and intersecting with themselves. This assumes that the arrow *i* that goes from X→X remains the same uniquely composed arrow throughout the entire process of its looping.

In other words, the existence of arrow *i* assumes that *i = i* and that it will not change when it loops and intersects with itself. However, each arrow's identity must be secured by yet another arrow, and so on indefinitely. But if we never reach the end or ground of arrows, we can only *assume* that *i* actually equals *i*. What if *i* was the product of an indefinite, metastable and non-well-defined composition of monomorphisms? In this case, the arrow is not a unique morphism, and the object (X) it describes through folding becomes indeterminate. The point I am making here is that we can reinterpret categories as being much more like 'quasi-categories' or 'kinetic categories' that are more ambiguous, mobile and ungrounded than classical categories.

In classical category theory, identity is explicitly defined as a *product* of arrows 'such that for every X, there is an arrow *i*: X→X such that for any arrow *a* going into or out of X, the composition of *a* with *i* (either *ia* or *ai*) is *a* itself'.[40] In other words, identity is the product of folding when an arrow loops back over itself and becomes the source and target of its own arrow. However, the number of arrows that go into composing the identity of that arrow is *indeterminate*. There could always be more arrows composing any arrow. This is what Joyal means by the term 'infinitely composed'. Therefore, the unique identity of any arrow in category theory is an assumption and not a foundation.[41]

This is where the idea of kinetic quasi-categories comes in as a more general framework to describe the existence of indeterminate objects, such as those in quantum physics. The four consequences of this kinetic operator are as strange as they are significant.

First, the kinetic operator means that the so-called identity of the object itself is only a tiny region of a much vaster world of relations. All relations have relations between their relations, and so on indeterminately. If an arrow's identity is secured by another arrow, then there is no absolute limit to the arrows that could all be considered 'equivalent' in a given arrow. But if every arrow is multiple, where does the identity of the arrow come from?

The mathematical term for this ambiguous kind of quasi-object in category theory is 'weak equivalence'. It means that the potentially vast relations and compositions equivalent to a given arrow are not or cannot be completely specified. We saw in the previous chapter how a very similar phenomenon occurred in Feynman's diagrams. There are always more diagrams below the current level where an indeterminate amount of smaller movements could have created the current diagram. But if the arrows of category theory are non-identical, they do not *strictly* determine the relations between objects.

My interpretation of this issue in category theory is that *movement* and *morphism* are irreducible to static topological shapes. There exists a deeper dimension of all mathematical objects that one cannot unambiguously treat as an object. The process of morphism itself is not static but in motion. This is why I think it would be useful to have a particular term in category theory, like the kinetic quasi-object, to describe an object whose relations are morphologically indeterminate.

We do not have to treat arrows in category theory as static. By 'static' I mean that they persist on their own in a single act by a mathematician. Instead, we can treat them as changing indeterminate relations sustained by ongoing 'metastable' mathematical actions. However, indeterminacy is not the same as non-identity. Indeterminacy is neither x nor *not x*. It does not follow the law of non-contradiction or the law of excluded middle. These laws are conventions of mathematics that we do not necessarily have to accept.

If an arrow is indeterminate, as I am interpreting it here through the idea of the quasi-object, it *may or may not be* equivalent to how it is being used in any given composition. This means that the objects created by arrows are also indeterminate. The objects of category theory are iterated or 'folded' from looping arrows. And these arrows are composed of looping arrows. Only through continual iteration and composition does indeterminacy become relatively metastable or 'weakly equivalent', without ever becoming strictly 'identical'.

The kinetic operator thus transforms arrows and objects into quasi-objects in a higher or kinetic category theory. The arrow and object are not static but are moving patterns of indeterminate process. The relation between objects is also changed when they become regions of quasi-arrow flows. So instead of a morphism *among* discrete objects, we have an indeterminate *change in the morphism itself*. This is the case because arrows are infinitely composed as described above. This is why I am calling it *'kinetic'* and not 'topological', since what is indeterminate is the movement of morphism itself.

The second important consequence of the kinetic operator is that the relations between arrows can become so vast that describing them runs up against the limits of computer technology and mathematicians' ability to diagram them. Indeterminate categories take time and movement to calculate. My point is not just that all mathematics requires the work of a mathematician. The point is that mathematicians' choices and labours are relevant to what one can mathematically say about quasi-objects. How many layers deep one goes in the categorical description of objects shapes the diagram in a meaningful way. Certain n-dimensional categories, for example, would take a mathematician a lifetime to describe. Others might take a computer years to describe. Others still might never be described. In a higher kinos category, the mathematical process shapes but is not identical to the mathematical product – unlike traditional categories.[42]

The third consequence is that every arrow is composed of a multiplicity of other arrows that we can 'treat as if'[43] they were isomorphic. Here classical category theory is correct to say that arrows are identical 'up to isomorphism', meaning that any arrow can be indefinitely composed and still function 'as identical' in a diagram. However, one cannot strictly guarantee this isomorphism. One can only practically use it 'as if' an indeterminate multiplicity of arrows were equivalent to 'an' arrow: f.

The fourth and most profound consequence of the arrow's indeterminacy is that it alters the first axiom of category theory: composition. The axiom of composition says that 'for any two arrows f and g that meet "head–to–tail" (that is, the same object is the target of f and the source of g), there is exactly one arrow called the "composition" of f and g that goes from the source of f to the target of g.'[44]

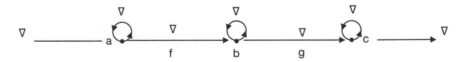

Figure 17.3 Category theory indeterminate composition. Author's drawing.

However, it is not necessarily true that there is only 'one arrow' in a composition, since any arrow might be composed of an indefinite number of other arrows and objects. If arrows and objects are only metastable processes, then composition only occurs 'as if' the arrow were equivalent or quasi–identical to its own compositions. This, as we argued above, can only be performed practically and never exhaustively shown. Even the axiom of choice fails here for reasons I discussed in prior chapters. In other words, there is a more primary indeterminacy to composition that we could formulate in the following way from a movement-oriented perspective:

An arrow is a kinetic operation, which is fundamentally indeterminate in direction and number. Through mathematical practice, one deploys this indeterminacy as a metastable isomorphism.

Fold

The second site of kinetic indeterminacy in category theory is its objects. Category theory treats objects as nodes of relations and products of looping arrows. This is consistent with the kinetic theory of objects as folds of flows.

If arrows are indeterminate, then so are objects. If an object is indeterminate, then one might create it from multiple internally differentiated arrows. However, since the number of arrows and their relations is indeterminate, they might include transformations that we are not aware of. If we dig deep enough beneath the object's layers, it may be otherwise than we think. Defining an object does fundamentally determine it. That is an insight of quasi-objects. A single object, *a*, may be indeterminately multiple and internally variable 'up to isomorphism'.

In other words, the traditional categorical object designators (*a*, *b*, *c*) cannot be made entirely determinate. Determinate objects are products of more primary indeterminate operators or morphisms. Since every categorical object is defined not by a static, self-identical arrow but by endless iterations of loops or folds, then objects are internally indeterminate. By interpreting the arrow and object as indeterminate, the result is quite different than in formalist 'logics of indeterminacy' or 'vague objects', in which an indeterminate object is treated as either a 'possible' object or as having some determinate logical operator.[45]

Category theory is fascinating because all its logical operators are indeterminate arrows that one can treat as metastable equivalences. One can treat indeterminate arrows as determinate operations such as negation $\neg$, equality $=$ or existence $\exists$. We can, therefore, interpret categorical objects in the following way:

For every object created by an arrow, there is an indeterminate identity of the object. Through mathematical practice, one deploys this indeterminacy as a metastable pattern of motion.

Field

The third site of kinetic indeterminacy in category theory is the empty category. Category theory treats all mathematical and logical structures initially as 'empty categories' open to axiomatic additions and transformations. In my kinetic interpretation, the empty category is also indeterminate.

The empty category contains no arrows and no objects. Without any arrows, the empty category sounds like the elastic background of homogeneous space put forward in modern analytic geometry and quantum theory. The categorical empty 'space' sounds as if it were entirely a priori. However, if diagrammatic arrows are performatively identical with mathematical existence, then there must also be some activity more primary than the presupposed background existence of the so-called 'empty category'. The empty category also has to be constructed.

There is no ultimate empty category in higher category theory. Every empty category might be inside another empty category. Since empty categories have no arrows or objects, there could be an indeterminate number of them lurking. So just as there are 'higher' 'quasi-categories', there also higher quasi-empty categories.[46] The existence of quasi-categories means that there is no absolute highest category or pre-given ground of all categories. This means that all empty categories are indeterminately grounded. In my view, one must immanently construct each empty category from the indeterminacy of empty quasi-categories. In other words:

Every empty category is the product of an indeterminate kinetic operation. A category is not merely assumed as a background but is actively constructed and produced without an ultimate foundation.

Conclusion

In this chapter, I have highlighted three significant aspects of category theory that are consistent with my kinetic philosophy of objects. In the next and final chapter, I propose a kinetic interpretation of another crucial contribution to a new theory of objects from chaos theory.

Notes

1. Jean-Michel Salanskis, 'Les mathématiques chez x avec x = Alain Badiou', in C. Ramond (ed.), *Alain Badiou: Penser le multiple* (Paris: Harmattan, 2002), 102; David Corfield, *Towards a Philosophy of Real Mathematics* (Cambridge: Cambridge University Press, 2003), 198.
2. Corfield, *Towards a Philosophy of Real Mathematics*, 198.
3. 'Work on the latter was initiated by Henri Poincaré (b.1854–1912) as a project to help develop tools to study differential equations qualitatively. Category theory is much closer to Riemann's mathematical program than to set theory. In fact, the fields of algebraic topology and differential geometry, from which category theory emerged, developed in the wake of Riemann's work

and were greatly influenced by it.' Simon Duffy, *Deleuze and the History of Mathematics: In Defence of the 'New'* (London: Bloomsbury, 2014), 154.

4. For more on Skolem's paradox, see Timothy Bays, *Reflections on Skolem's Paradox* (Ann Arbor: UMI Dissertation Services, 2002).

5. Formal systems such as propositional logic can be consistent and complete in the way Godel proves set theory cannot, and other such axiomatic systems cannot.

6. A. S. Troelstra, 'History of Constructivism in the 20th Century', available at <http://citeseerx.ist.psu.edu/viewdoc/download?doi=10.1.1.77.926&rep=rep1&type=pdf> (last accessed 4 March 2021).

7. Colin McLarty, 'Two Constructivist Aspects of Category Theory', *Philosophia Scientiæ*, 6 (2006), 95–114, available at <http://citeseerx.ist.psu.edu/viewdoc/download?doi=10.1.1.532.2855&rep=rep1&type=pdf> (last accessed 4 March 2021)

8. Alain Badiou, *Mathematics of the Transcendental*, trans. A. J. Bartlett and Alex Ling (London: Bloomsbury, 2017), 15.

9. Badiou, *Mathematics of the Transcendental*, 1.

10. Fernando Zalamea, *Synthetic Philosophy of Contemporary Mathematics* (Falmouth: Urbanomic, 2012), 60; Albert Lautman, *Mathematics, Ideas, and the Physical Real*, trans. Simon Duffy (London: Bloomsbury, 2011), xxviii.

11. 'Basically, when he maintains that it is necessary to accept "the legitimacy of a theory of abstract structures, independent of the objects connected together by these structures", Lautman is very close to a mathematical theory oriented towards the structural relations beyond objects: the mathematical theory of categories.' Lautman, *Mathematics, Ideas, and the Physical Real*, xxxv.

12. Lautman, *Mathematics, Ideas, and the Physical Real*, xxxiv.

13. 'The theory proved to hold such universal power that it developed into an entire mathematical ontology.' Badiou, *Mathematics of the Transcendental*, 13.

14. Badiou, *Mathematics of the Transcendental*, 55.

15. Badiou, *Mathematics of the Transcendental*, 56.

16. Rocco Gangle, *Diagrammatic Immanence: Category Theory and Philosophy* (Edinburgh: Edinburgh University Press, 2016), 6.

17. See Geoffrey Hellman, 'Mathematical Pluralism: The Case of Smooth Infinitesimal Analysis', *Journal of Philosophical Logic*, 35 (2006), 621–51, available at <https://link.springer.com/content/pdf/10.1007%2Fs10992-006-9028-9.pdf> (last accessed 4 March 2021). However, non-contradiction will frequently be maintained in category theory because once A has been constructed and so is proved true, ~A can then be inferred to be false.

18. Badiou, *Mathematics of the Transcendental*, 13, my italics.

19. Badiou, *Mathematics of the Transcendental*, 14.

20. Badiou, *Mathematics of the Transcendental*, 15.

21. Saunders Mac Lane, *Categories for the Working Mathematician* (New York: Springer, 1998), 9.

22. Zalamea, *Synthetic Philosophy of Contemporary Mathematics*, 60.

23. Zalamea, *Synthetic Philosophy of Contemporary Mathematics*, 295–6.

24. Zalamea, *Synthetic Philosophy of Contemporary Mathematics*, 295–6.
25. See 'La pensée mathématique', in J. Cavaillès, *Oeuvres complètes de philosophie des sciences* (Paris: Hermann, 1994), 593–630.
26. Zalamea, *Synthetic Philosophy of Contemporary Mathematics*, 271.
27. Zalamea, *Synthetic Philosophy of Contemporary Mathematics*, 272.
28. 'The processual, nonstatic Plato, a Plato not fixed to a reification of the idea, a Plato whom Natorp recuperated at the beginning of the twentieth century, and to whom Lautman and Badiou would later return, seems to constitute the nondual, mobile base that mathematics requires: an apparent contradiction in terms – for the approaches customary to analytic philosophy of mathematics, the "base" should not turn out to be mobile.' Zalamea, *Synthetic Philosophy of Contemporary Mathematics*, 329.
29. Lee Smolin, *Three Roads to Quantum Gravity* (New York: Basic Books, 2017), 30.
30. Smolin, *Three Roads to Quantum Gravity*, 30–1.
31. Zalamea, *Synthetic Philosophy of Contemporary Mathematics*, 293.
32. Badiou, *Mathematics of the Transcendental*, 14.
33. Cited in Zalamea, *Synthetic Philosophy of Contemporary Mathematics*, 190.
34. Badiou, *Mathematics of the Transcendental*, 13, my italics.
35. Zalamea, *Synthetic Philosophy of Contemporary Mathematics*, 332.
36. Zalamea, *Synthetic Philosophy of Contemporary Mathematics*, 275.
37. Zalamea, *Synthetic Philosophy of Contemporary Mathematics*, 182, 121, 346.
38. These are based explicitly on 'Schrödinger logics' in quantum physics. Erwin Schrödinger argued that identity lacks sense for elementary particles of modern physics.
39. André Joyal, *The Theory of Quasi-Categories and its Applications*, 202, available at <http://mat.uab.cat/~kock/crm/hocat/advanced-course/Quadern45-2.pdf> (last accessed 4 March 2021).
40. Gangle, *Diagrammatic Immanence*, 89.
41. This is what is called 'transfinite composition'. 'It is a means to talk about morphisms in a category that behave as if they were the result of composing infinitely many morphisms.' See <https://ncatlab.org/nlab/show/transfinite+composition> (last accessed 4 March 2021).
42. This point is much deeper than Mac Lane's theory of objects as 'sets varying in time'. The kinetic indeterminacy of the arrow itself undergirds the conditions of discrete temporal variation as such. Objects are pure indeterminate variation. See Mac Lane, *Categories for the Working Mathematician*, 402–4: 'Functions between pairs of indexed sets, e.g., X0 and X1; Y0 and Y1; as the objects, with certain pairs of functions running from the Xi and the Yi, as the morphisms.'
43. 'Notions of homotopy are perhaps more common than one might expect since the philosophy of model categories shows that simply specifying a class of "weak equivalences" in a category, a collection of morphisms which we wish to treat as if they were isomorphisms, produces a notion of homotopy.' Omar Antolín Camarena, 'A Whirlwind Tour of the World of (∞,1)-categories', 1, available at <https://arxiv.org/pdf/1303.4669.pdf> (last accessed 4 March 2021).

44. Gangle, *Diagrammatic Immanence*, 89.
45. P. F. Gibbins, 'The Strange Modal Logic of Indeterminacy', *Logique et Analyse*, NS, 25.100 (1982), 443–6, available at <http://www.jstor.org/stable/44084046> (last accessed 4 March 2021).
46. Clark Barwick, Emanuele Dotto, Saul Glasman, Denis Nardin and Jay Shah, 'Parametrized Higher Category Theory and Higher Algebra: Exposé I – Elements of Parametrized Higher Category Theory', 4, available at <https://www.maths.ed.ac.uk/~cbarwick/papers/basics.pdf> (last accessed 4 March 2021).

18

The Contemporary Object III:
Chaos Theory

The third contribution to the loop theory of the object emerged in the 1970s with the study of non-linear and complex dynamical systems, or 'chaos theory'. For more than 200 years, Newtonian physics provided the foundations for all macro-level physical processes. Even statistical mechanics was merely a way to deal with our ignorance concerning large numbers of presumably Newtonian processes, such as atoms bouncing around in gases. The assumption was that objects all followed general equations whose initial conditions could be known.

However, chaos theory showed that tiny differences in the initial conditions of how an object began moving or how someone measured it could create widely diverging outputs over time. Each object seems to follow a unique evolution that renders its precise prediction impossible. However, the exact trajectory for objects is impossible to know because the precise initial conditions are impossible to know. In my view, this is an essential lesson for a new theory of the object.

In this chapter, I want to propose an *indeterministic* interpretation of chaos theory. Some chaos theory interpretations say that there is an objective set of initial conditions, but we cannot know them. In my view, however, we cannot know the precise initial conditions of any object *in principle*, due in part to quantum indeterminacy and entanglement. Therefore, I see no reason to assume an initial objective state or a deterministic movement of objects from that state. Instead, I believe that chaos theory teaches us that each object has a singular trajectory that we cannot know in advance but that *tends* to produce dynamic patterns over time.

Studying these patterns of motion across different macroscopic scales can help us understand physical systems' common features, even if it cannot predict them precisely. In this chapter, I hope to show that interpreting the kinetic indeterminacy at the heart of the initial conditions of things is entirely consistent with and even beneficial for chaos theory.

What is chaos theory?

One of the earliest precursors of chaos theory was Henri Poincaré. In the 1880s Poincaré tried to apply Newton's 'two-body' equations that had successfully modelled the orbit of the earth and the moon around their joint centre of gravity to a 'three-body' equation, which included the sun.[1] In the two-body equation, you can treat one body as static and calculate only the other body moving around it as a closed system, like the moon's rotation around the earth. The result is an ellipse. However, once you add a third body to the equation, you can no longer treat any body as a fixed reference point, since each of the bodies' changing gravity continually affects the other two bodies. These tiny changes have enormous consequences over time. The equation, therefore, had to be continuously recalculated without end. Poincaré discovered that a final solution was impossible because small changes in input do not result in small changes in outputs. The equations required to model the three-body problem were 'non-linear'.

Non-linear equations differ from linear ones. In mathematics, linear equations are ones where the inputs' addition equals the sum of the outputs. For example, $y = x$, $1 + 2 = 1 + 2$. These are 'linear' equations because you get a straight line when you map them on a cartesian plane. 'Non-linear' equations are ones where the sum of the inputs does not equal the outputs. For example, $y = x^2$. If x is 3, y is 9 not ($3 + 2 = 5$). When you graph non-linear equations, you get curves. If a variable in your equation has an exponent greater than 1, then small input changes do not always result in small changes in outputs.

When one calculates non-linear equations, one cannot easily extrapolate the path that an object will travel based on how it has previously travelled. The equation has to be calculated again and again with each variation in the physical conditions. This was too labour-intensive for earlier mathematicians and physicists, and so they tended to favour linear approximations and treated certain processes such as turbulence and friction as random.

However, with computers in the 1970s, non-linear equations could be rapidly calculated and graphed with millions of iterations. To the great surprise of physicists, what appeared to be random results at first began to produce distinct patterns over time. For example, the French astronomer Michel Hénon used a computer to calculate planetary orbits around a galactic centre through millions of iterations. What at first appeared to be random movements turned out to be periodic patterns that never repeated precisely. In chaos theory, these patterns are called 'attractors'.

The American meteorologist Edward Lorenz discovered similar patterns while monitoring weather simulations on a computer.[2] By mistakenly changing a tiny number in his initial equations, Lorenz discovered that extremely divergent weather patterns resulted. The Polish-born French-American mathematician Benoit Mandelbrot found similar recurring patterns at every scale in data on cotton prices.[3] Mandelbrot called these new kinds of objects 'fractals'. A fractal is an object whose irregularity is constant or 'self-similar' over different scales. Trees are fractal because their larger limbs' branching pattern repeats in their smaller limbs and their leaves' veins. Other examples include the Koch snowflake, the Menger sponge, river deltas and lightning.[4]

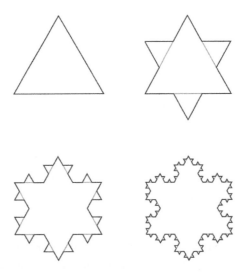

Figure 18.1 Koch snowflake. https://en.wikipedia.org/wiki/Koch_snowflake#/media/File:KochFlake.svg.

Figure 18.2 Menger sponge. https://en.wikipedia.org/wiki/Menger_sponge#/media/File:Menger-Schwamm-farbig.png.

Figure 18.3 Tree fractal. https://en.wikipedia.org/wiki/Fractal#/media/File: Simple_Fractals.png.

Figure 18.4 Lightning. https://en.wikipedia.org/wiki/Lightning_injury#/media/ File:Lightning3.jpg.

The Australian physicist Robert May found similar periodic patterns in population biology.

> May found that the time it took for the system to oscillate back to its starting point doubled at certain critical values of the equation [see figure 18.6]. Then after several period-doubling cycles, the insect population in his model varied randomly, just like real insect populations, showing no predicable period for return to tis original state.[5]

The American mathematician James Yorke found that if you map non-linear equations over many iterations, they go through three stages that lead to a fourth 'chaotic' phase.[6] The American physicist Robert Shaw even found periodic attractor patterns in the drips from a water tap. One might think that the interval between drips would be random, but 'as more and more points were laid down on the graph, a shape emerged from the mist that looked remarkably like the cross section of a strange attractor known as the Hénon attractor'.[7]

Beyond merely periodic attractors, the Belgian-French physicist David Ruelle further described what he called a 'strange' attractor in fluid turbulence patterns that occurs after a series of three periodic attractors. This is what Yorke had called, 'Period three implies chaos.' Instead of the third pattern merely increasing its dimensions indefinitely, Ruelle argued that the third pattern 'itself begins to break apart! Its surface enters a space of (fractional) dimensions' vaguely surrounding the attractor.[8]

Amid much scepticism from classical and quantum physicists, these new chaos theorists found one another's work across different disciplines. They formally announced the emergence of their new science in December 1977 at their first conference. One of the most critical developments that made such a conference and science possible was the American physicist and mathematician Mitchell Feigenbaum's discovery of a universal mathematical pattern common to all these chaotic phenomena. As one of the conference organisers, Joseph Ford, said, 'Mitch had seen universality and found out how it scaled and worked out a way of getting to chaos that was intuitively appealing. It was the first time we had a clear model that everybody could understand.'[9]

In the sections below, I want to offer an indeterministic interpretation of three aspects of chaos theory: its hybridity, pedesis and feedback.

Hybridity

Chaos theory is not a foundational or universal science that tries to explain everything. It is not a theory of the quantum world or all mathematical logic. Instead, it is a hybrid kind of science of the macroscopic world. It studies the emergence of patterns across extremely heterogeneous phenomena in numerous sciences.

In chaos theory, there are no universal laws that exactly predict the singular behaviour of objects. Instead, chaos theory describes the *tendency* of macroscopic processes to move in a series of patterns over time. Mitchell Feigenbaum offered the first theoretical unification of this process when he identified the period-doubling ratios shared by all chaotic processes.

> Feigenbaum showed that the fine details of these different systems don't matter, that period doubling is a common factor in the way order breaks down into chaos. He was able to calculate a few universal numbers representing the ratios in the scale of transition points during the doubling process. He found that when a system works on itself again and again, it will exhibit change at precisely these universal points along the the scale.[10]

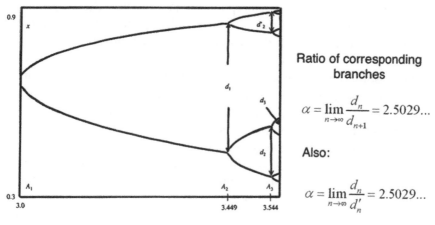

Ratio of corresponding branches

$$\alpha = \lim_{n \to \infty} \frac{d_n}{d_{n+1}} = 2.5029...$$

Also:

$$\alpha = \lim_{n \to \infty} \frac{d_n}{d'_n} = 2.5029...$$

Figure 18.5 Feigenbaum constant. Author's drawing.

Feigenbaum identified a general tendency in the evolution of processes. However, precise measurements of the initial conditions of objects are still impossible. Where precisely will a process bifurcate into different patterns? It all depends on the precision of the initial conditions. Feigenbaum represented this sensitivity in the mathematical form of a limit approaching infinity.

Feigenbaum borrowed the idea of 'renormalisation' used by quantum field theorists in the 1950s to 'collapse infinities into manageable quantities'.[11] Quantum field theorists, in turn, borrowed the idea of spontaneous 'symmetry breaking' or period doubling from chaos theory.[12] The practice of renormalisation allowed both sciences to make a determinate measurement of an indeterminate process and get usable results, even though classical and quantum indeterminacy are different.

Once a finite bifurcation point is experimentally measured, Feigenbaum's mathematical scaling follows a constant ratio with the rest of the period–doubling cycles. In this way, 'Feigenbaum's constant' worked in much the same way as the standard model did in quantum physics in the 1970s. It provided an elegant and straightforward quantitative (renormalised) formula that scientists from numerous other disciplines could understand and apply in their areas. 'Armed with these numbers and the knowledge of period doubling, scientists all over the world soon began finding chaos everywhere.'[13]

Chaotic objects are not fixed essences or relations of resemblance between fixed objects, but rather dynamic patterns of motion.[14] Chaos theory treats objects as processes of iteration and not as extrapolations of fundamental particles. As Feigenbaum writes,

> There's a fundamental presumption in physics that the way you understand the world is that you keep isolating its ingredients until you understand the stuff that you think is truly fundamental. Then you presume that the other things you don't understand are details. The assumption is that there are a small number of principles that you can discern by looking at things in their pure state – this is the true analytic notion – and then somehow you put these together in more complicated ways when you want to solve more dirty problems. [Chaos theory, however,] is a general description of what happens in a large variety of systems when things work on themselves again and again. It requires a different way of thinking about the problem.[15]

In this way, chaos theory functions as a hybrid, process-oriented description of many kinds of different objects. Scientists have applied the patterns of chaos theory to almost every scientific discipline, including meteorology, anthropology, sociology, physics, environmental science, computer science, engineering, economics, biology, ecology, philosophy, astrophysics, information theory, computational neuroscience, psychology, and many others.

Pedesis

The chaotic object is pedetic or indeterminate. This indeterminacy is related to but different from quantum and categorical indeterminacy. Chaotic indeterminacy is the impossibility of measuring the initial conditions of any *macroscopic* object with absolute precision. In my view, this is not because there are objective initial conditions that we cannot describe. It's because the initial conditions are, at the smallest level, related to quantum indeterminacy, and at the broadest level are related to the globally entangled universe, including the observer.

Furthermore, there are no *determinate* initial conditions in chaos theory because we are not dealing with discrete, static or non-relational things. We are dealing with macroscopic kinetic *processes*. This is why they are so 'sensitive'. This is also why there are no fixed probability ranges in chaos theory. The trajectory of each process-object is singular and relational.

The chaotic object's pedesis has four main interrelated features: 1) sensitivity to initial conditions, 2) progressive determination, 3) irreversibility and 4) non-locality.

Initial conditions

Chaotic objects are sensitive to their initial conditions because there is no fixed, determinate set of initial conditions.[16] How one measures or describes an object in motion affects how the object will move and be measured in the future.

Chaotic objects are recursive and non-linear *processes* in which small differences feed back into the process and eventually diverge significantly from infinitesimally similar processes. This is why, when Lorenz entered .506 into his weather simulator instead of entering .506127 so as to save space, he found that small changes in the initial determination of the mathematical object had disproportionate long-term effects. Each determination of an 'initial condition' produces singular and *fundamentally* unpredictable long-term consequences. Instead of the modernist dream of a totality of deterministic outcomes or probabilities, chaos theory contributes to a different theory of the object. What is vital about chaos theory, in my interpretation, is that we cannot know the initial conditions and the totality of possible outcomes with absolute certainty because every trajectory is singular.

As the Russian chemist Ilya Prigogine and the Belgian philosopher Isabelle Stengers write,

> It is true that there is still a trajectory description if initial conditions are known with infinite precision. But this does not correspond to any realistic situation. Whenever we perform an experiment, whether by computer or some other means, we are dealing with situations in which the initial conditions are given with a finite precision and lead, for chaotic systems, to a breaking of time symmetry.[17]

The initial conditions of an object are not a totality of discrete and finite states. Objects are emergent metastable processes in non-local relations. The initial conditions of objects are *not reducible to a determinate or discrete set of objects*. Prigogine and Stengers argue that 'The initial conditions of a single trajectory correspond to an infinite set $\{u_n\}$ ($n = -\infty$ to $+\infty$). But in the real world, we can only look through a finite window. This means that we are able to control an arbitrary but limited number of digits u_n.'[18]

In my view, however, we should not interpret this 'infinity' as a set of discrete states, but instead as an indeterminate process. An event's probability is not $1/\infty$ (.0 repeating) because infinity is not the same as indeterminacy. Eventually, a close look at the initial conditions will reach the vacuum fluctuations of quantum physics. Chaos science progressively shapes indeterminate movements into metastable objects.

Progressive determination

Since there is no objective state of initial conditions, only processes, objects are progressively determined. There is no total set of pre-existing possible states if there are no initial conditions but only singular trajectories. As Manuel DeLanda writes,

> Experimentalists progressively discern what is relevant and what is not in a given experiment. In other words, the distribution of the important and the unimportant defining an experimental problem (what degrees of freedom matter, what disturbances do not make a difference) are not grasped at a glance the way one is supposed to grasp an essence (or a clear and distinct idea), but slowly brought to light as the assemblage stabilizes itself through the mutual accommodation of its heterogeneous components . . . affects of the experimentalist's body are meshed with those of machines, models and material processes in order for learning to occur and for embodied expertise to accumulate.[19]

We cannot know the nature of the object in advance of its unfolding as a process. In the initial conditions and at each 'step', the object's movement follows an indeterminate and yet progressively metastable path. Measurement is not a one-time affair that deduces all trajectories or probabilities. 'We need to conceive a *continuum* which yields, through progressive differentiation, all the discontinuous individuals that populate the actual world.'[20]

This method is similar to the one the Russian mathematician Andrey Markov used. Each new motion of the object is not determined by any previous motion but only indeterminately related to the one immediately preceding it. The finite probabilities that Markov said result from the chain (of determinate sets and objects) can emerge only based on the more pedetic movement of the object itself.

Irreversibility

Motion is irreversible because it is pedetic or indeterminate. If the movement of each object is truly singular, then it cannot be reversible. This is a key thermodynamic insight of chaos theory. Reversibility, according to Prigogine, is an idealistic fantasy of scientific mathematisation. Matter and motion can only obey the same laws forwards as backwards if they are passive and uncreative. Chaos theory takes seriously the second law of thermodynamics, which implies an extremely high tendency towards entropy.[21]

Irreversibility is not just a side effect of subjective human observation. We see entropy because we live in an entropic universe. Entropy is something nature does to itself. As Prigogine writes, 'Theoretical reversibility arises from the use of idealizations in classical or quantum mechanics that go beyond the possibilities of measurement performed with any finite precision. The irreversibility that we observe is a feature of theories that take proper account of the nature and limitation of observation.'[22]

Mathematical and physical reversibility is a perspective from nowhere and assumes perfect knowledge. However, there is no perfect knowledge and no total objective situation to know. The French polymath Pierre-Simon Laplace imagined that if there were a demon that knew the position and momentum of all matter in the universe, it would know their past and future with perfect deterministic accuracy. In my view, chaos theory rejects this idea *in principle*. We cannot know the movement of matter in advance of its irreversible progressive determination. The universe is capable of real novelty and singularity because it is indeterministic. There are no laws that absolutely determine the movement of matter before its unfolding. Every object has its singular 'Lyapunov time', named after the Russian mathematician Aleksandr Lyapunov, or the 'time after which its periodicity becomes genuinely chaotic'.

If there are no determinate initial conditions, then there are no determinate final conditions either. There are only truly emergent processes and patterns. In my interpretation of chaos theory, indeterminacy is not an epistemological inability but a real process of irreversible entropy.

Non-locality

The initial conditions of the chaotic object are indeterminate in part because they are also non-local. The trajectory of an object in motion is not determinate or deterministic because it is related to larger processes called 'Poin-caré resonances' within which it is only one region.[23] As Prigogine writes,

> resonances are not local events inasmuch as they do not occur at a given point or instant. They imply a nonlocal description and therefore cannot be included in the trajectory description associated with New-tonian dynamics. As we shall see, they lead to diffusive motion . . . similar to that of the 'random walk', or 'Brownian motion'.[24]

'Transient interactions' are relations that occur only once and which scientists describe by localised distribution functions. However, 'persistent interactions' continue to evolve and require delocalised distribution func-tions, as in atmospheric meteorology. The pedesis of the chaotic object is due to these persistent interactions with non-local processes. For example, the intervalic periodicity of drips of water from a tap in Shaw's experiment cannot be reduced to a purely local or global deterministic cause. Each drip is part of a non-local flow of water pressure and movement through exter-nal pipes and other water usage activities elsewhere in the city.

Local patterns are regions of broader non-local processes. 'Matter', as Prigogine and Stengers write, 'acquires new properties when far from equilibrium in that fluctuations and instabilities now are the norm. Matter becomes more "active".'[25] Local matters become more active and creative when they are persistently interacting with broader non-local processes. The '"indeterministic hypothesis" is the natural outcome of the modern theory of instability and chaos'.[26]

Local matters become more active and creative when they are persis-tently interacting with broader non-local processes. We can thus distin-guish between two kinds of physical indeterminacy. At the classical level of chaotic objects, propagating with a finite speed above the speed of light, we cannot make exact predictions because the system is too globally com-plex, relational and too sensitive to the initial conditions. This is a classical indeterminacy. At the quantum level, objects are indeterminate due to the fluctuation of the fields that compose them. Since quantum systems are immanent with classical ones, it would seem to follow that they have some non-trivial relationship. Prigogine and others have argued for the continu-ity between these two forms of indeterminacy, now called 'quantum chaos theory', but research is still ongoing.[27]

Feedback

The second prominent feature of the chaotic object is its fundamentally relational and periodic nature. The chaotic object is not defined primarily by centripetal, centrifugal, tensional or even elastic patterns of motion. Instead, all of these motions are only regionally stable formations or periodic patterns produced by matter's pedetic movement. The chaotic object is not an essence or an eternally stable structure. It is a loop, a fold, or periodicity in the movement of matter. Objects are attractors, or series of synchronised attractors, whose regional stability is the effect of a tendency towards chaotic or pedetic motion. The chaotic object emerges and dissipates because of its pedetic motion, not despite it.

Periodicity

Chaotic objects are processes that emerge from 'dense periodic orbits'. No matter how different two patterns of motion are, they will eventually enter into a periodic pattern of regional stability as they interact. In chaos theory, this is called 'coupled oscillation'. As two connected moving systems begin to interact, they create a mutual transformation or metastable synchrony. For example, when Christiaan Huygens set two pendulum clocks with different timings on the same wall of his room, he noticed that over time the two clocks began to synchronise their swings through the vibrations of the wall and floor. The American mathematician Steven Strogatz has written extensively on the application of cyclical periodicity and spontaneous synchrony in biology, economics, quantum physics, astrophysics, hydraulics and other areas. He writes,

> At the heart of the universe is a steady, insistent beat: the sound of cycles in sync. It pervades nature at every scale from the nucleus to the cosmos. Every night along the tidal rivers of Malaysia, thousands of fireflies congregate in the mangroves and flash in unison, without any leader or cue from the environment. Trillions of electrons march in lockstep in a superconductor, enabling electricity to flow through it with zero resistance. In the solar system, gravitational synchrony can eject huge boulders out of the asteroid belt and toward Earth; the cataclysmic impact of one such meteor is thought to have killed the dinosaurs. Even our bodies are symphonies of rhythm, kept alive by the relentless, coordinated firing of thousands of pacemaker cells in our hearts. In every case, these feats of synchrony occur spontaneously almost as if nature has an eerie yearning for order.[28]

Periodicity is the spontaneous generation of rhythmic coordination in nature and is the foundation for the coordination that defines the practice of object-making in human science. Periodic coordination is not unique to humans.

Chaotic objects are loops, intervals or pleats in matter. The history of science has privileged four major patterns of motion. Chaos theory describes each of these four patterns as 'periodic attractors'. The 'point attractor' describes centripetal motion, the 'limit cycle' describes centrifugal motion, the 'torus attractor' describes tensional and elastic motion, and the 'chaotic attractor' describes pedetic motion. Let's look more closely at each one in turn.

Point attractor

A point attractor emerges when objects move towards a single point. It is a centripetal motion that heads towards a single 'basin of attraction'. For example, a sinkhole in the earth is a basin of attraction that centripetally draws in rainwater. A swinging pendulum moves back and forth, slowed by gravity and dynamic friction from the air until it comes to rest at its lowest energy point.

Limit cycle

If something continually pushes the pendulum forwards, such as a clock's weight mechanism, the pendulum will tend towards a resting point *and*

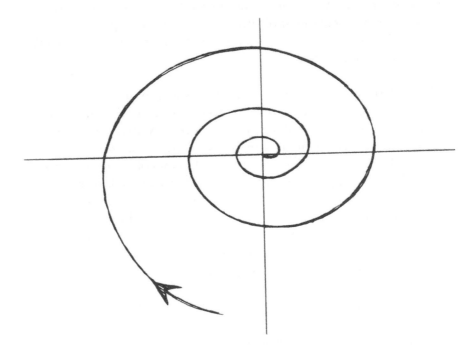

Figure 18.6 Point attractor. From John P. Briggs and F. D. Peat, *Turbulent Mirror: An Illustrated Guide to Chaos Theory and the Science of Wholeness* (New York: Harper and Row, 2000), 36.

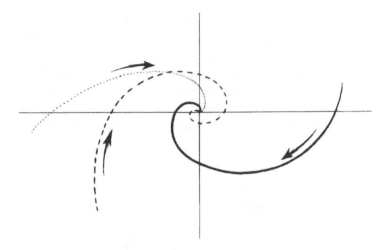

Figure 18.7 Phase space of a pendulum. From John P. Briggs and F. D. Peat, *Turbulent Mirror: An Illustrated Guide to Chaos Theory and the Science of Wholeness* (New York: Harper and Row, 2000), 36.

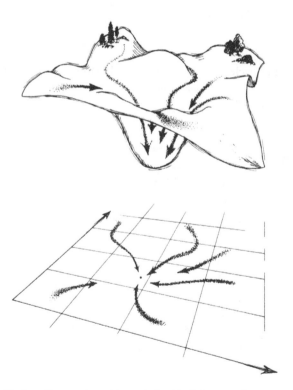

Figure 18.8 Basin of attraction. From John P. Briggs and F. D. Peat, *Turbulent Mirror: An Illustrated Guide to Chaos Theory and the Science of Wholeness* (New York: Harper and Row, 2000), 36.

move centrifugally back out. 'Rather than the pendulum being attracted to a fixed point, it is drawn toward a cyclical path in phase space. This path is called a limit cycle, or limit cycle attractor.'[29]

Such attractors are surprisingly resistant to perturbations of their motion because of the dynamic feedback between entropic viscosity and the input of external energy. Something similar occurs in predator–prey cycles. 1) A food source such as trout becomes abundant, 2) then a predator population

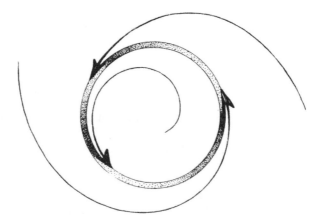

Figure 18.9 Limit cycle attractor. From John P. Briggs and F. D. Peat, *Turbulent Mirror: An Illustrated Guide to Chaos Theory and the Science of Wholeness* (New York: Harper and Row, 2000), 37.

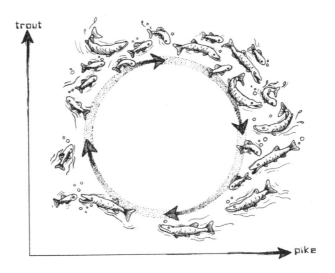

Figure 18.10 Predator–prey feedback cycle. From John P. Briggs and F. D. Peat, *Turbulent Mirror: An Illustrated Guide to Chaos Theory and the Science of Wholeness* (New York: Harper and Row, 2000), 38.

such as pike increases and eats all the food, 3) but dies off when the food becomes scarce. 4) Eventually, the food source regenerates, and the cycle repeats but never in exactly the same way. Limit cycles are not perfect circles but emergent objects from non-linear pedetic processes.

Torus attractor

A torus attractor occurs when the periodicity of a limit cycle has at least two degrees of freedom. This produces a tensional motion between the two coupled oscillators. For example, if we loosened the pendulum's suspension system such that it moved back and forth and side to side, its range of motion or 'phase space' would look like a torus or 'doughnut shape'. If these coupled oscillations move together in a simple or classical ratio where one is twice as big as the other, the twists around the torus will join up exactly.

However, suppose the frequencies are not precise, simple ratios such as 1/2, 1/3 or 1/4, but are irrational, that is, fractions or ratios resulting in infinite, non-repeating decimal expressions such as √2. In that case, the motion never quite repeats along the same path of the torus. Its motion is 'quasi-periodic'. Instead of a held tensional proportion, the system takes on an elastic motion, expanding and contracting within the torus system's metastable parameters. Quasi-periodicity shows how iterated singular trajectories can create a metastable oscillatory system.

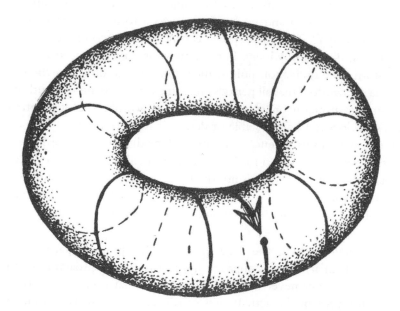

Figure 18.11 Rational torus attractor. From John P. Briggs and F. D. Peat, *Turbulent Mirror: An Illustrated Guide to Chaos Theory and the Science of Wholeness* (New York: Harper and Row, 2000), 40.

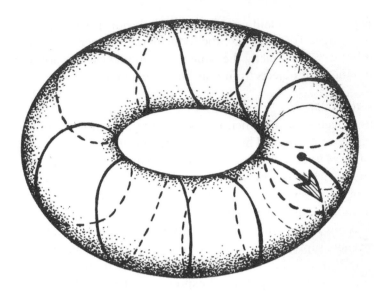

Figure 18.12 Irrational torus attractor. From John P. Briggs and F. D. Peat, *Turbulent Mirror: An Illustrated Guide to Chaos Theory and the Science of Wholeness* (New York: Harper and Row, 2000), 41.

Strange attractor

A strange attractor occurs when the object's movement becomes completely non-repeating and no longer follows a torus or multidimensional torus pattern at all. The phase space becomes entirely composed of singular trajectories, diverging widely. For example, as a river's speed increases in a rainstorm, the oscillation points, including the raindrops on the surface, become so manifold that all periods and attractors break up into turbulence. The strange attractor or 'chaotic attractor' is not like the other attractors because it does not have a stable or determinate periodicity.

However, this turbulence is not merely the dissolution of order and stable attractors. If it goes on for long enough, it can stabilise into the attractors described above. Chaos is not the failure of order but its source. Chaos or pedesis is immanent to patterns of motion, but it can become more or less stabilised for a while.[30] As Manuel DeLanda writes,

> As is well known, the trajectories in this [phase] space always approach an attractor asymptotically, that is, they approach it indefinitely close but never reach it. This means that unlike trajectories, which represent the actual states of objects in the world, attractors are never actualized, since no point of a trajectory ever reaches the attractor itself. It is in this sense that singularities represent only the long-term tendencies of a system, never its actual states.[31]

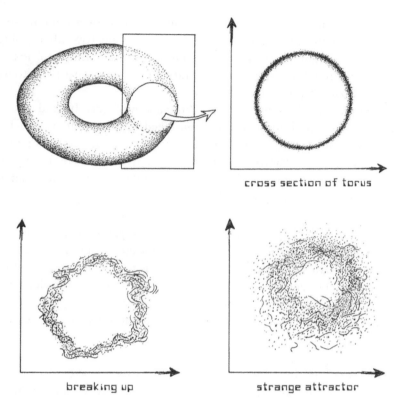

cross section of torus

breaking up

strange attractor

Figure 18.13 Strange attractor. From John P. Briggs and F. D. Peat, *Turbulent Mirror: An Illustrated Guide to Chaos Theory and the Science of Wholeness* (New York: Harper and Row, 2000), 46.

Topological mixing

This brings us to the third prominent feature of the chaotic object. Topological mixing is the tendency for any close periodic pattern of motion to diverge from all patterns. This is a crucial point. The chaotic object is not a single type of attractor but rather the continual transformation from one attractor to another. The chaotic object's pedetic movement allows it to enter into loop patterns and bifurcate into other loops and unravel all these loops. 'If the perturbation is large enough, a distribution of attractors may cease to be structurally stable and change or bifurcate into a different one. Such a bifurcation event is defined as a continuous deformation of one vector field into another topologically inequivalent one through a structural instability.'[32] Imagine smoke rings slowly dissolving into the air. Over time, their metastable pattern increasingly mixes with the surrounding air and unravels into tiny wisps of smoke.

Topological mixing has significant consequences for a contemporary loop theory of the object. If all chaotic objects tend towards topological mixing, turbulence and chaotic attractors at some point, feedback and resonance are not limited to any arbitrary finite phase space or 'topos'. Feedback is part of a non-local process.

This requires a different way of thinking about objects. Instead of treating them as discrete, bounded things, we need to treat them as non-local processes passing through various attractors and mixing back into pedesis. In other words, the chaotic object's feedback is not just something that occurs between pre-existing objects but is a holistic feature of entangled physical systems. As Prigogine and Stengers write,

> Classical dynamics extracts a given number of particles and considers their motion in isolation; irreversibility occurs when interactions never cease. In short, dynamics corresponds to a reductionist point of view in the sense that we consider a finite number of molecules in isolation. Irreversibility emerges from a more holistic approach in which we consider systems driven by a large number of particles as a whole.[33]

In chaos theory, there are no objects in isolation. There are only loop objects that fold up and dissipate as regions of more widely entangled processes.

Kinetic chaos

The chaotic object is not reducible to a discrete unit or fixed trajectory. It is instead a process that loops, folds and cycles around attractors. These attractors then synchronise, couple and eventually become turbulent. In the final section of this chapter, I show how my kinetic theory of the object is consistent with the indeterministic interpretation of chaos theory developed above.

Flow

The initial conditions of chaos theory are indeterminate flows of matter in motion. In a sense, though, they are neither 'initial' nor 'conditions'. They are not initial because there is no absolute beginning of motion. They are not conditions because there are no discrete determinate states that can predict the future with certainty.

Experimental measurements intervene and 'renormalise' this indeterminate flow. They make 'agential cuts',[34] as Barad calls them, by localising the initial conditions in some finite initial measurement. Prigogine and Stengers are explicit about the nature of this constitutive measurement act when they say that all 'terms that tend toward infinity are meaningless'. So Prigogine and Stengers 'eliminate these divergences in [their] statistical description'.[35] There is no such thing as infinity in nature, but chaos theories such as Feigenbaum's use mathematical infinities in their formalisms to stand in for the unobjectifiable and indeterminately sensitive flow of matter, which they must cut somewhere to get finite answers.[36]

This is why Prigogine and Stengers describe the performative practice of science as a 'dialogue with nature'.[37] Science constitutively co-produces a field of objects based on a particular measurement of an indeterminate, non-objective flow. This is one of the key theses I have tried to develop in this book: the basis of objects is not an object. Science does not discover pre-existing objects but weaves and folds them from flows. This is why nature responds as an immanent agent of scientific practice.

In my view, if chaos theory wants to keep the dialogue with nature open, it has to acknowledge the non-objective conditions of objects and the role of scientists as co-constitutive agents in the creation of these sensitive and responsive initial conditions. This is what I call the kinetic operator.

An experimental measurement is not like a slice or cut out of a whole. It is a process of transformation because measurement and coordination contribute to the dissipation of nature. Kinetic operators progressively determine but do not exhaust the initial conditions. The initial conditions are not a total phase space of possible starting points. Scientists weave their model of phase space from the flow of indeterminate initial conditions.

Fold

Chaotic objects are metastable folds in indeterminate flows. As they cycle and iterate, their broader non-local relationships alter them. These small changes over time can create enormous consequences and new patterns of motion. In my interpretation, this is not 'deterministic chaos'. Although the non-linear equations and algorithms used by chaos scientists are deterministic, this does not mean that they are genuinely dealing with deterministic processes. The same revealing mathematical results of long-term unpredictability are consistent with an indeterministic interpretation. Indeterminacy is the source of singularity and unpredictability. Chaos theory renormalises the indeterminacy of the initial conditions and non-local resonances and provides deterministic approximations.

In my view, using computers to map non-linear equations over time is a beautiful and revealing way to show just how singular any determinate initial condition is. It is a much better approximation of objects in motion than classical non-linear approximations. But nature does not look *exactly* like the equations of chaos theory predict it to look, because nature does not start with determinate renormalised initial conditions. Nature's folds are much more indeterminate. For example, there is a striking difference between the natural fractal patterns we seen in ferns and the exact fractal patterns derived from mathematical algorithms that look like ferns. The two look more similar than non-linear approximations, but the natural fern is different because it is indeterminately relational while the mathematical one is deterministic. The deterministic algorithms of chaos theory alter their content with each iteration, but nature continually alters everything.

Field

Chaotic objects emerge and move across phase spaces. In chaos theory, a phase space is a range of possible actions within which attractors emerge, diverge and mix topologically. This phase space is consistent with what I call a 'field' in my kinetic interpretation. A field is not a total range of discrete states but a metastable region woven from initially indeterminate conditions. The use of probability in chaos theory is based on the renormalisation of these indeterminate and sensitive initial conditions into a regional phase space. If chaos theory treats objects as discrete probable states, it does so only by intervening or weaving them from a wider, non-local field of relations. Science is not the language of nature but the co-creation of a field of coordinated objects.

The movement of chaotic objects proceeds by progressive determinations towards an unpredictable future without complete knowledge of its past or initial conditions. However, probabilistic mathematical models of this process are explicitly renormalised. For example, a Markov probability chain assumes 1) a closed set of possible options, 2) a strictly linear determination of motion, and 3) the assumption of discrete states for isolated objects. All three assume that a discrete measurement has renormalised the indeterminacy of the initial conditions and their non-local resonances.

Every probability field, amplitude or distribution, even a progressively determined one, still requires the field's constructive determination by a kinetic operator. The kinetic operator is not separate from the initial conditions but co-constitutes them.

Contemporary theories of the object show us that the conditions of objects are not other objects but indeterminate processes. This does not mean that science should give up on weaving and ordering objects; rather, it should affirm its co-creative vocation. This is how I interpret the kinetic meaning of renormalisation. It is an explicit acknowledgement of the non-objective conditions of objects and the creative act of scientific labour. Objectification is not about total knowledge of pre-existing discrete objects, but the proliferation and deployment of new objects woven from indeterminate motion. The object is immanent to the kinetic process of coordination that produces it.

If the kinetic operator is continuous with the object's construction, then it is also the *real immanent construction of nature*. In this sense, objects are not reductions of nature, and subjects are not separate from nature. They *are* nature. Objects are progressive and metastable local determinations of fundamentally non-local processes. This is what I have tried to emphasise with the idea of a loop object.

Conclusion

Quantum, category and chaos theory all contribute insights into a new theory of the object. In my view, these theories acknowledge, to some degree, the priority of indeterminate movement. However, they also include hybrid aspects of all four previous theories of objects. In particular, the modern probabilistic theory of the object is still the dominant interpretation among practising scientists.

In these last few chapters, the key idea I wanted to share was that these three recent sciences point beyond the four theories of the object discussed in this book. They all directly confront the primacy of indeterminate motion even if they have also found ways to renormalise it into effective equations.

I aimed to show that the world, as far as we know, is not only objects, and that objects are not static or discrete. Science co-creates and weaves metastable objects from indeterminate flows of matter. This does not mean that nature is whatever humans say it is. Nature and humans *really* work together in scientific practice. Renormalisation techniques can be practical and effective as creative approximations, but metaphysical beliefs in determinism, causality, randomness and probability are unnecessary and potentially dangerous to the creative efforts of science because they try to explain away the creative indeterminacy of matter instead of working within it.

I used the term 'kinetic operator' as a way to name this generative source of objects and the processes of their metastable orders, whether they are made by humans or nature or both. Finally, my argument was that the kinetic theory of the object is consistent with recent scientific evidence and offers an interpretation that I hope might keep science open and sensitive to the primacy of indeterminate motion in nature and away from metaphysical attempts to explain it away.

Notes

1. Jules Henri Poincaré, 'Sur le problème des trois corps et les équations de la dynamique. Divergence des séries de M. Lindstedt', *Acta Mathematica*, 13 (1890), 1–270.
2. Edward N. Lorenz, 'Deterministic Non-periodic Flow', *Journal of the Atmospheric Sciences*, 20.2 (1963), 130–41, Bibcode: 1963JAtS...20..130L.
3. Benoît Mandelbrot, 'The Variation of Certain Speculative Prices', *Journal of Business*, 36.4 (1963), 394–419.
4. Benoît Mandelbrot, *The Fractal Geometry of Nature* (New York: Macmillan, 1982).
5. John P. Briggs and F. D. Peat, *Turbulent Mirror: An Illustrated Guide to Chaos Theory and the Science of Wholeness* (New York: Harper and Row, 2000), 60.
6. T. Y. Li and J. A. Yorke, 'Period Three Implies Chaos', *American Mathematical Monthly*, 82.10 (1975), 985–92.
7. Briggs and Peat, *Turbulent Mirror*, 88.
8. Briggs and Peat, *Turbulent Mirror*, 51. Ruelle's theory is experimentally supported by the work of Harry Swinney and Jerry Gollub.
9. James Gleick, *Chaos: Making a New Science* (London: The Folio Society, 2015), 184.
10. Briggs and Peat, *Turbulent Mirror*, 64.
11. Gleick, *Chaos*, 179.
12. Manuel DeLanda, *Intensive Science and Virtual Philosophy* (New York: Bloomsbury, 2013), 18–19.
13. Briggs and Peat, *Turbulent Mirror*, 64.
14. DeLanda, *Intensive Science and Virtual Philosophy*, 176, n. 57.
15. Gleick, *Chaos*, 185.
16. Robert Devaney, *An Introduction to Chaotic Dynamical Systems* (Boulder: Chapman and Hall/CRC, 2018).
17. Ilya Prigogine and Isabelle Stengers, *The End of Certainty: Time, Chaos, and the New Laws of Nature* (New York: The Free Press, 1997), 105.
18. Prigogine and Stengers, *The End of Certainty*, 101.
19. DeLanda, *Intensive Science and Virtual Philosophy*, 172.
20. DeLanda, *Intensive Science and Virtual Philosophy*, 72.
21. 'Nature involves both time-reversible and time-irreversible processes, but it is fair to say that irreversible processes are the rule and reversible processes the exception. Reversible processes correspond to idealizations: We have to ignore friction to make the pendulum move reversibly. Such idealizations are problematic because there is no absolute void in nature. As previously mentioned, time-reversible processes are described by equations of motion, which are invariant with respect to time inversion, as is the case in Newton's equation in classical mechanics or Schrodinger's equation in quantum mechanics. For irreversible processes, however, we need a description that breaks time symmetry.' Prigogine and Stengers, *The End of Certainty*, 18.
22. Ilya Prigogine, *From Being to Becoming: Time and Complexity in the Physical Sciences* (San Francisco: W. H. Freeman, 1980), 215.

23. Prigogine and Stengers, *The End of Certainty*, 112.
24. Prigogine and Stengers, *The End of Certainty*, 42.
25. Prigogine and Stengers, *The End of Certainty*, 65.
26. Prigogine and Stengers, *The End of Certainty*, 56.
27. 'The physical reasons for which we have to leave Hilbert space are related to the problem of persistent interactions mentioned above, which requires a holistic, nonlocal description. It is only outside Hilbert space that the equivalence between individual and statistical description is irrevocably broken, and irreversibility is incorporated into the laws of nature.' Prigogine and Stengers, *The End of Certainty*, 97.
28. Steven Strogatz, *Sync: The Emerging Science of Spontaneous Order* (London: Penguin, 2004), 1
29. Briggs and Peat, *Turbulent Mirror*, 37.
30. As the first-century Roman poet Lucretius believed, indeterminate swerving generates order. See Thomas Nail, *Lucretius I: An Ontology of Motion* (Edinburgh: Edinburgh University Press, 2018).
31. DeLanda, *Intensive Science and Virtual Philosophy*, 23.
32. DeLanda, *Intensive Science and Virtual Philosophy*, 23.
33. Prigogine and Stengers, *The End of Certainty*, 114.
34. Karen Barad, *Meeting the Universe Halfway: Quantum Physics and the Entanglement of Matter and Meaning* (Durham, NC: Duke University Press, 2007), 148.
35. Prigogine and Stengers, *The End of Certainty*, 112.
36. I disagree with Prigogine and Stengers that these flows are 'meaningless'. I would say rather that nature is the immanent act of meaning-making.
37. Prigogine and Stengers, *The End of Certainty*, 153.

Conclusion

Today, hybrid objects of all kinds surround us. The proliferation and vast circulation of objects worldwide is forcing us to rethink the nature and relationship between movement and objects. In particular, the contemporary sciences of quantum, category and chaos theory bring us face to face with the indeterminate movements at the heart of objects. This indeterminacy is the source of objectivity and its limits.

In this book, I have aimed to make three main contributions to the philosophy, history and science of objects.

Part I: the kinetic object

My first contribution was to propose that we think about objects as processes of movement instead of static or discrete entities. To do this, I put forward the concepts of flow, fold and field to help us think about how processes might be woven into metastable patterns, like threads into a piece of fabric. The goal was to show how we could explain the relatively stable appearance and reproducibility of objects from more primary processes. I offered a process-based theory of numbers, knowledge, scientific experiment and observation based on this idea. Using this conceptual framework I developed a process theory of science and knowledge as the creation and distribution of objects-in-motion.

Part II: history of the object

The second contribution was that I used Part I's conceptual framework to reinterpret the history of scientific knowledge from prehistory to the present. From this history, I wanted to show that certain humans invented four main kinds of objects following four unique patterns of motion.

I showed that during prehistory humans invented ordinal objects following a centripetal pattern of motion. In the ancient world, they invented cardinal objects that had a centrifugal motion. In the medieval and early modern world, they invented intensive objects using a tensional pattern of motion. Finally, scientists invented what I called potential objects following an elastic pattern of motion in the modern world. In each of these historical epochs, a new kind of object rose to dominance. All the older ones persisted and mixed with them. These historical chapters' main contribution was to show that there is not only one kind of object and that objects are emergent properties of material history, not unchanging forms.

Part III: the contemporary object

The third contribution of my theory of the object was to offer an interpretation of three key contemporary sciences that pointed towards a fifth kind of 'indeterminate' object. I also argued that contemporary objects have aspects of all four types of historical objects.

If we want to understand contemporary objects, we have to see how they constitutively mix with the history of objects that came before. In particular, I aimed to show that my kinetic theory of the object was consistent with these recent sciences based on their indeterministic interpretation. These chapters' main contribution was to warn against metaphysical interpretations of contemporary science as deterministic, random and probabilistic, and offer an alternative interpretation. Science, I argued, is a co-creative act with nature that shapes indeterminate processes into metastable objects and patterns we call knowledge.

Limitations

I mentioned some of the book's limitations in the introduction, but I want to mention a few more here. First, I want to be clear that all kinds of processes create objects, not just humans or scientists. The universe creates all kinds of objects that have nothing to do with humans. This book is limited to the history of objects created by science.

Second, and perhaps most importantly, this book is limited strictly to studying objects' *kinetic structures*. The theory of objects in this book is not a complete history of science or every great scientist in the Western tradition. It does not pretend to do biographical, comparative or systematic justice to Western science. What is unique about *Theory of the Object* is its focus on the previously hidden kinetic structures operating within the history of knowledge. They reveal a subterranean science of motion and an agency of objects. This kinetic and process history of science has been systematically marginalised and occluded by the history of solids, statics, linear equations and particles. The science of static objects has prevailed. I have limited my inquiry to motion not to be reductionistic, but to provide an alternative historical lineage leading up to the sciences' present state.

Third, *Theory of the Object* describes each of these dominant fields of objects separately as they emerge historically. In reality, though, all the fields coexist and mix to some degree or other. To show all such mixtures and degrees for each historical period is too large a task and must be reserved for future studies. Therefore, this book considers only the dominant distributions of objectification during the period of their historical rise to prominence.

Fourth, the present work is limited to a particular geo-historical lineage from prehistory, to the Near East, and into Euro-Western modern scientific practice. In no way does this suggest that the West has the only or the best sciences. On the contrary, the revelation of the primacy of motion at the heart of Western science is my way of undoing certain prevailing notions of science as the progressive study of fixed essences and laws of nature.

By showing the secret material-kinetic conditions of this history, I hope to have challenged the foundationalism of science and knowledge. This book is restricted geographically purely because of the practical limitations of length and the linguistic and cultural limitations of its author, not because of any notion of Western superiority.

Future directions

In response to these limitations, there are also several areas where future work could be done. First, one could expand the project's historical scope beyond the human sciences by looking at the structure of knowledge in other living and non-living beings more broadly. The creation of objects is not limited to human beings.

Second, the kinetic theory of objects could be a useful supplement to other theories of objects interested in considering the material and kinetic aspects of things.

Third, one could show how the kinetic indeterminacy at the heart of objects has always played a role in destabilising and supporting each historical epoch's dominant regimes.

Fourth, one could expand the geographical scope of this kinetic theory to Eastern and colonial worlds. It is hard to imagine that Western science would be what it is without the influence and exploitation of these other spheres. While the history of Western science in this book is not, I hope, technically inaccurate, it is by no means the whole story – none ever is. The Eastern and colonial worlds have their own major and minor historical periods and arts that proceeded alongside those of the West, influencing and deriving influence from it.

Fifth, I was often forced to sacrifice empirical depth for historical breadth in this book. But I couldn't have both without a much larger book or stricter limitations. As such, I had to leave out many great scientists and kinds of knowledge. Even the sciences I have managed to include have been cut short. This book was twice as long it is now before numerous rounds of editing. One could further develop several sections in more detail following the kinetic method introduced here.

Above all, I hope the reader will find this book, or some parts of it, helpful enough to be applied elsewhere. Hopefully, this is only the beginning of a more sustained inquiry into the kinetic study of objects.

Index

Page numbers in *italics* refer to figures, and those with the suffix 'n' refer to notes (e.g. 15n).